CODE OF ESTIMATING PRACTICE

Endorsed by the Estimating Practice Committee of The Chartered Institute of Building

Acknowledgements

The first edition of the Code of Estimating Practice was produced by the Institute's Estimating Practice Committee in April 1966 and this has been revised on three occasions since that time.

The Code has now been totally rewritten and updated to incorporate many important new features which are currently regarded as an integral part of good estimating practice.

The rewriting was carried out by D. T. Stubbington, *F,* who was supported by a Steering Group composed of representatives from the CIOB's Estimating Practice Committee. The composition of the Steering Group was,

Chairman	K. B. Jones, MCIOB
	A. W. J. Appleton BEd, FRICS, FCIOB
	A. P. Frost, MCIOB
	D. G. Potter, FCIOB, FBIM
	D. T. Stubbington, FCIOB, FBIM, FFB
Secretary	G. P. Cottrell, BSc MCIOB, Head of Practice

The author and Steering Group wish to acknowledge the considerable and invaluable contribution made by G. P. Cottrell in the preparation of this Code.

The Code of Estimating Practice has been approved and endorsed by the Estimating Practice Committee.

D. G. Potter,
Chairman of the Estimating Practice Committee of
The Chartered Institute of Building

Contents

Acknowledgements	iii
1.00 **TERMINOLOGY**	1
2.00 **INTRODUCTION**	5
3.00 **PRESELECTION**	9
3.01 Definition	9
3.02 Marketing objectives	9
3.03 Updating lists of approved contractors	9
3.04 Objectives of preselection	10
3.05 Programme for preselection of contractors	10
3.06 Preselection information from consultants	10
3.07 Analysis and response by the contractor	12
3.08 Analysis of the contractor's response	13
4.00 **DECISION TO TENDER**	19
4.01 Introduction	19
4.02 Receipt of tender documents	19
4.03 Projects for which preselection has not occurred	20
4.04 Projects where preselection has occurred	20
4.05 Inspection of the tender documents	20
4.06 Information required for the preparation of an estimate	21
4.07 Conditions of contract	23
4.08 Workload and timetable	23
4.09 Type of work and resources needed	23
4.10 Tender information	24
4.11 Management decision	25
5.00 **PROJECT APPRECIATION**	27
5.01 Management of the estimate	27
5.02 Ensure all tender documents received	27
5.03 Timetable for production of the estimate and tender	29
5.04 Thorough examination of the tender documents	29
5.05 Prime Cost Sums, Provisional Sums daywork and contingencies	33
5.06 Method statement and tender programme	36
5.07 Site visit	41
5.08 Visit to consultants	42
6.00 **ENQUIRIES AND QUOTATIONS**	45
6.01 Preparation of documents for enquiry purposes	45
6.02 Enquiries	47
6.03 Quotation analysis	53
7.00 **ALL IN RATES AND UNIT RATES**	59
7.01 Summary of procedure for establishing 'all in rates'	59
7.02 Estimating 'all in rates' (Stage 1)	61
7.03 Example of calculation of an 'all in hourly rate' for labour	63
7.04 Receipt, analysis and selection of quotations	71
7.05 Establishment of net unit rates (Stage 2)	72
8.00 **ALTERNATIVE ESTIMATING TECHNIQUES AND DAYWORK**	81
8.01 Spot items	81
8.02 Operational estimating	81
8.03 Extension of bills of quantities	83
8.04 Prime Cost of Daywork	84
9.00 **COMPLETING THE COST ESTIMATE**	91
9.01 Nominated suppliers	91
9.02 Nominated sub-contractors	92
9.03 Late quotations	95
9.04 Project overheads (Stage 3)	95
9.05 Preliminaries and conditions of contract	110
9.06 Fluctuating price	110
9.07 Allowance for a firm price tender	113
9.08 Cash flow	113
9.09 Review and finalise the cost estimate	114
10.00 **ESTIMATOR'S REPORT AND ADJUDICATION**	117
10.01 Estimator's summary analysis and report	117
10.02 Adjudication	126
10.03 Submission of a tender	128
11.00 **ACTION AFTER SUBMISSION OF A TENDER**	129
11.01 Assessing tenders and notifying results	129
11.02 Adjustment of errors	129
11.03 Action with a successful tender	131
11.04 Action with an unsuccessful tender	132

Flow charts and forms

Preselection flow chart	8
Preliminary/tender enquiry form	14-15
Decision to tender flow chart	18
Project appreciation flow chart	26
Co-ordination meeting checklist	28
Abstract for materials enquiries	31
Abstract for sub-contractor enquiries	32
Schedule of PC and provisional sums daywork and contingencies	34
Site visit report	38-40
Enquiries and quotations flow chart	44
Materials quotations register	49
Plant quotations register	52
Domestic sub-contractors' register	54
'All-in-rates' and unit rates — alternative estimating techniques and daywork flow chart	58
Composition of net unit rates	60
Calculating the 'all-in-hourly rate'	72-73
Mechanical plant — calculating 'all-in-rates' per hour	78
Operational estimating	82
Specimen summary of account	88
Completing the cost estimate flow chart	90
Project overheads schedule	96-107
Estimator's reports and adjudication flow chart	116
Estimator's summary, analysis and report	118-125
Comparison of results	130

1.00 Terminology

The following are the meanings of the principal specialist terms used in this Code.

ADJUDICATION
The action taken by management to convert an estimate into a tender.

ALL-IN LABOUR RATE
A compounded rate which includes payments to operatives and the costs which arise directly from the employment of labour. Variable costs, such as travelling time or abnormal overtime, are normally excluded.

ALL-IN MATERIAL RATE
A rate which includes the cost of material delivered to site, waste, unloading, handling, storage and preparing for use.

ALL-IN MECHANICAL PLANT RATE
A compounded rate which includes the costs originating from the ownership or hire of plant together with operating costs.

APPROVED CONTRACTORS
Approved contractors are those who have demonstrated that they have the expertise, resources, ability and desire to tender for a proposed project. Selection of such contractors is normally by preselection procedures.

ATTENDANCE
The labour, plant, materials or other facilities provided by the main contractor for the benefit of the sub-contractor and for which the sub-contractor normally bears no cost.

BRIEF
The definition of a client's building requirements.

BUILDABILITY
The extent to which the design of a building facilitates ease of construction subject to the overall requirements for the completed building.

CONSULTANTS
The client's advisors on design, cost and other matters. Such advisors may include project managers, architects, engineers, quantity surveyors, accountants, bankers or other experts relating to the client's particular needs.

COST
The term cost, without qualification, means the estimated cost of the physical production of work.
Note: Estimated cost must not be confused with historical cost; historical cost is the cost of construction revealed only after the work has been executed.

COST RECORDS
Records of historical cost and notes of the conditions prevailing when such cost was incurred.

DOMESTIC SUB-CONTRACTORS
Sub-contractors selected and employed by the contractor.

DOWN TIME
The period of time that plant is not operating. This may be due to breakdown, servicing time or inability to operate due to external factors.

ESTIMATING
The technical process of predicting costs of construction.

ESTIMATOR
A person carrying out the estimating function in a building organisation. Such a person may be a specialist or he may carry out the estimating function in conjunction with other functions, such as quantity surveying, general management etc.

FIRM PRICE CONTRACT
A fixed price contract which does not allow for its prices to be adjusted for fluctuations.

FIXED PRICE CONTRACT
A price which is agreed and fixed before construction starts. It may or may not be a firm price contract.

FLUCTUATIONS
The increase or decrease in cost of labour, plant or materials which may occur during a contract.

GC/WORKS/1
The standard form of contract used by Central Government for all major building and civil engineering works.

GANG COST
A grouping of labour costs to include principal and supporting labour associated with a particular trade.

HEAD OFFICE OVERHEADS
The cost of administering a company and providing off-site services. The apportionment of head office overheads to projects and to the company as a whole is decided by management as part of management policy.

JCT '63
The Standard Form of Building Contract 1963 Edition prepared by the Joint Contracts Tribunal.

JCT '80
The Standard Form of Building Contract 1980 Edition prepared by the Joint Contracts Tribunal.

LABOUR ONLY SUB-CONTRACTORS
Sub-contractors whose services are limited to the provision of labour.

LUMP SUM CONTRACT
A fixed price contract where the contractor has undertaken to be responsible for executing the whole of the contract work for a stated total sum.

MANAGEMENT
The carrying out of the function of general management and the responsibility for making the decision to tender and for adjudicating the tender.

MATERIAL
Material or components.

MARK UP
The sum added to an estimate in respect of head office overheads and profit.

METHOD STATEMENT
A statement of the construction methods and resources to be employed in executing construction work. This statement is normally closely linked to the tender programme.

NJCC
The National Joint Consultative Committee for Building.

NOMINATED SUB-CONTRACTOR/SUPPLIER
A sub-contractor/supplier who is selected by the architect or engineer and not by the contractor.

OPEN COMPETITION
The method of selecting tenderers and obtaining tenders by a general invitation to contractors who apply for tender documents. The number of tenderers is not usually limited and reputation and ability to execute the work satisfactorily are not always considered.

PREQUALIFICATION
The provision by a contractor of information as part of a pre-selection process.

PRESELECTION
The establishment of a list of contractors with suitable experience, resources, ability and desire to execute a project, bearing in mind the character, size, location and timing of the project.

PRIME COST (PC)
When used in bills of quantities and specifications prime cost means work to be carried out by nominated sub-contractors or suppliers of materials and goods. The contractor may also be invited to carry out work covered by a PC sum in certain instances.

PROJECT OVERHEADS
The cost of administering a project and providing general plant, site staff, facilities and site-based services and other items not included in all in rates.

PROVISIONAL SUM
A sum included in bills of quantities for work anticipated but insufficiently designed or detailed to permit descriptions and measurement in accordance with the requirement of the Standard Method of Measurement (SMM).

SELECTIVE TENDERING
A method of selecting tenderers and obtaining tenders whereby the number of contractors invited to tender is limited to the inclusion of contractors who are considered suitable and able to carry out the work. This suitability is usually determined by preselection procedures.

SELECT COMPETITION
The endeavour to gain a contract against a limited number of competitors by the submission of a tender.

STANDING PLANT
Plant retained on site which is not working but which the contractor is still liable to pay for.

TENDER
The sum of money, time and other conditions required by the tenderers to carry out the specified building work.

TENDERING
A separate and subsequent commercial function based upon the estimate.

TENDER DOCUMENTS
Documents submitted for the information of tenderers.

TENDER PROGRAMME
The initial version of the master construction programme prepared during the tendering period, to enable the main contractor to appreciate the important time and resource considerations of the project in terms suitable to the preparation and submission of his tender.

TENDER TIMETABLE
The timetable for the preparation of the estimate, all necessary supporting actions and for the subsequent conversion of the estimate into a tender and submission by the date stipulated in the tender documents.

WORKING RULE AGREEMENT (WRA)
National Working Rules for the building industry produced by the National Joint Council for the Building Industry.

2.00 Introduction

This Code of Practice is intended to provide an authoritative guide to good practice in estimating for building work from preselection to the acceptance of a successful tender. It will be of use to students, estimators and those members of the building team who are involved with the production and use of an estimate.

The Code distinguishes between *Estimating* and *Tendering* and considers the formal steps to be taken in converting the net cost estimate into a tender at the adjudication stage.

> *Estimating* is defined as the technical process of predicting costs of construction. *Tendering* is a separate and subsequent commercial function based upon the net cost estimate.

The basic principles of estimating are set out in this document, which also considers the integration of estimating and management and links with the functions of programming, buying and construction.

Management is an important aspect in the production of the estimate and the subsequent tender. The estimator must have management responsibility within the estimating department and a responsibility for managing the production of the estimate. This responsibility requires the estimator to undertake various actions in managing resources and information during the tendering period.

This management role must not be underestimated. The estimator has to ensure that other company departments and staff work to his requirements, produce information on time and in the format required, and that effective operating procedures and lines of communication are established between departments to allow the efficient production of the estimate.

The initiation and control of enquiries, quotations and programme are important aspects of the estimating process. The estimator must co-ordinate any temporary works design and liaise with general management on construction method and available resources. In many organisations the estimator also has a responsibility for tendering and the subsequent management of the project in the construction phase if the tender is successful.

An estimate must be prepared in a way that is explicit and consistent, and which takes account of methods of construction and all circumstances which may affect the execution of the work on the project. A sound estimate can only be achieved when each operation or item is analysed into its simplest elements and the cost estimated methodically on the basis of factual information.

The basic principles of estimating described in this Code will be suitable for wide application by those estimating for various types of building work. In specific circumstances the procedures may be modified or simplified but the essential sequence of tasks and method in which information is obtained is unlikely to change. The use of ordered and logical methods of estimating will provide management with information on the future demands on resources. The estimate, in conjunction with the tender programme and method statement, provides a means of measuring the financial consequences of any delays that may occur during the execution of the project. Comparisons can be made between estimated and actual levels of productivity, allowing management to take the necessary action when required.

The estimate is, therefore, a base document which acts as an important control to construction and on which many assessments and judgements will be made during the construction phase.

This Code examines the various stages in the production of the estimate and its conversion to a tender. The various functional needs of the estimator are considered and procedures described to establish how the basic components of the estimate are identified and costs of the resources to be employed on a project are established.

Attention is drawn to the practice of establishing initially a net cost estimate. Project overheads are then determined, taking note of the method statement and tender programme. An important aspect of the estimator's work is to identify areas of risk and matters which may affect the costs of a project, for consideration by management at the adjudication stage. The Code illustrates such matters and demonstrates how they are considered at the adjudication stage.

The text assumes that bills of quantity have been measured in accordance with the Standard Method of Measurement of Building Works (SMM).

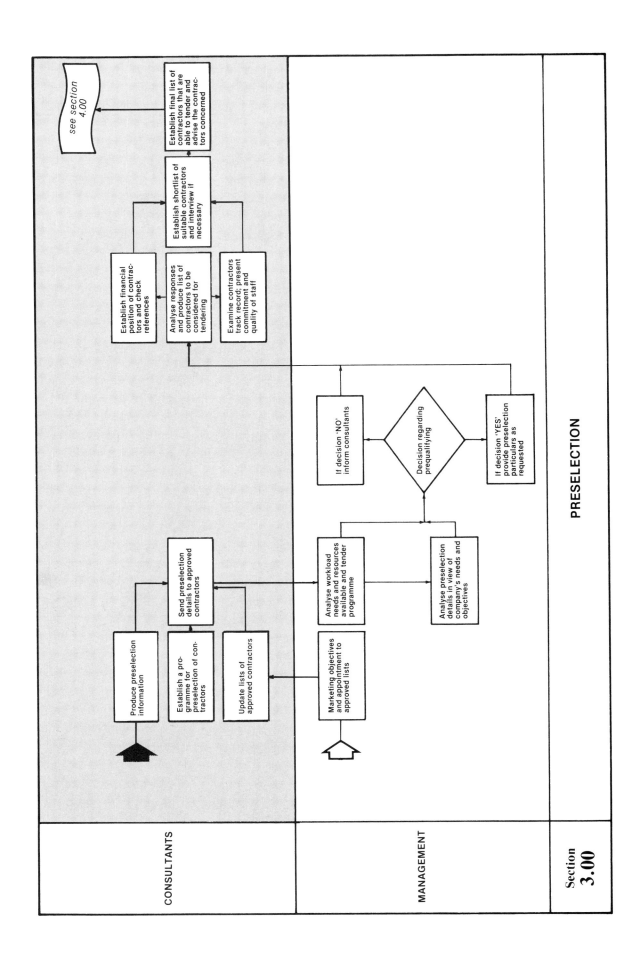

3.00 Preselection

3.01	DEFINITION	3.06	PRESELECTION INFORMATION FROM CONSULTANTS
3.02	MARKETING OBJECTIVES		– *Basic information*
			Client
3.03	UPDATING LISTS OF APPROVED CONTRACTORS		Project details
			Form of contract and contract particulars
3.04	OBJECTIVES OF PRESELECTION		– *Changes in information*
			– *Client's questionnaire to be completed by the contractor*
3.05	PROGRAMME FOR PRESELECTION OF CONTRACTORS	3.07	ANALYSIS AND RESPONSE BY THE CONTRACTOR
	3.08	ANALYSIS OF THE CONTRACTOR'S RESPONSE	
		– *Preliminary information*	
		– *Interview*	
		– *Selected list*	
	ADDITIONAL READING		

3.01 DEFINITION

Preselection is concerned with the establishment of a list of contractors with suitable experience, resources, ability and desire to execute a proposed project, bearing in mind the character, size, location and timing of the project.

Final selection is usually made either by negotiation or tender on the whole or part of the works, once the list of contractors has been determined by preselection procedures.

3.02 MARKETING OBJECTIVES

The contractor, from his marketing objectives, will establish the type, size and location of projects for which he wishes and is competent to tender. To be successful these marketing objectives must be translated into firm enquiries to tender. Contractors will seek to achieve this by being placed on lists of approved contractors retained by client organisations who are engaged or proposing to build.

(The subject of marketing is outside the scope of this Code).

3.03 UPDATING LISTS OF APPROVED CONTRACTORS

It is important that clients and clients' representatives monitor and regularly update their lists of approved contractors to:

- exclude companies whose performance has been unsatisfactory;
- introduce suitable new companies which can demonstrate the required qualities and abilities;
- compile the lists in a form appropriate to the class of project.

3.04 OBJECTIVES OF PRESELECTION

At this stage the client and his consultants will be concerned with the identification of contractors with the necessary skills, resources and desire to carry out the works. The contractor will be concerned with planning the work load in his estimating department, satisfying himself that projects are suited to his particular skills, meet a requirement regarding his work load objectives and that he has adequate management and resources available to support the project at the time stated.

3.05 PROGRAMME FOR PRESELECTION OF CONTRACTORS

Preselection will occur some time before any formal invitations to tender are issued. Consultants must ensure that a realistic programme is produced covering the whole of the preselection and tendering period and that adequate time is given for each stage. On major works preselection may occur three months before tender documents are issued. In the case of smaller contracts a period of four to six weeks is more usual.

The contractor must be given adequate time to evaluate the information provided and to prepare the information and data required for preselection purposes. Comprehensive and detailed information will be required and on major works such submissions could take several weeks to prepare.

3.06 PRESELECTION INFORMATION FROM CONSULTANTS

Preselection data is usually obtained by means of a questionnaire sent by the client or consultant to various contractors. These contractors will be chosen:

- on a basis of local standing;
- by personal knowledge and experience of the client or consultants;
- by the use of approved lists;
- as a result of preliminary investigations made as to the contractor's ability to carry out the work;
- from previous experience of using the contractor;
- because of representations made by the contractor for inclusion on appropriate lists;
- from the contractor's response to published advertisements.

To make a realistic judgement on whether to tender for a project, a contractor must be given the following preliminary information:

Basic information

Client

- details of the client, or if a subsidiary company, details of the holding company;
- full particulars of consultants to be used on the project, including their duties and responsibilities;
- particulars of proposed site supervision to be provided by the client or consultants;
- a description of the tender documents, their expected date of issue, the period available for tendering, the acceptance period for the tender and time when unsuccessful tenderers will be notified;
- whether the project, either in its present or a different form, has been the subject

- of a previous invitation to tender;
- the latest date for receipt of acceptance of invitation to tender;
- the number of tenders to be invited.

Project details

- the location of the proposed works, including preliminary drawings and a site plan;
- description of the project;
- approximate cost of the project;
- the date for commencement of the contract;
- the period for completion;
- details of any phasing;
- an outline of the method of construction;
- access problems;
- special operational space requirements;
- sufficient dimensions and specification details to permit evaluation of the project;
- detairs of work to be carried out by nominated sub-contractors, approximate value, and names if known.

Form of contract and contract particulars

- form of contract to be used;
- provisions for liquidated damages;
- proposed amendments to standard forms of contract and appendix;
- details of interim payments;
- retention conditions;
- bonding requirements;
- details of insurance;
- provisions for fluctuations in cost.

Changes in information

The contractor must be given prompt notification of any changes in the information provided, both during the period that the invitiation is open and subsequently before the despatch of tender documents. It is important that consultants keep to the dates stated in pre-qualification documents regarding the issuing of tender documents and that the contractor should be kept informed of any changes in these dates.

Client's questionnaire to be completed by the contractor

In addition to providing this essential project information, clients and their consultants may well find it advantageous to produce a standard questionnaire for completion by contractors in order to obtain comparable relevant information concerning the resources, management expertise and experience. Such a questionnaire should seek to obtain the following information from the contractor:

- name of firm;
- registered office address;
- local office address and telephone number;
- date of formation or registration;
- registration number;
- if a member of a group of companies, the name and address of the parent company;
- whether the parent company guarantees the performance of its subsidiary in accordance with the proposed terms of contract;
- nominal and paid up share capital;

- particulars of insurances held by the contractor;
- annual turnover during the past five years and current workload;
- details of contracts of a similar nature and size in hand, or carried out during the last three years including:
  (i) description of project;
  (ii) client;
  (iii) architect;
  (iv) quantity surveyor;
  (v) engineer;
  (vi) value;
  (vii) contract period and information on actual completion date compared with programmed completion date.
- number of employees employed on a regular basis under the following headings:
  (i) administrative;
  (ii) technical;
  (iii) operatives;
- list of trades usually sub-contracted;
- name and addresses of referees from whom references may be obtained;
- names and qualifications of the directors and the proposed management team for the project;
- other comments that may be of assistance to the client in arriving at a selection decision, including method of financial and programme control, organisation structure and preliminary ideas on method of construction;
- the person dealing with the application in the contractor's organisation.

Clients will wish to be assured of the financial stability of the contractors who will be invited to prequalify. Most contractors will provide bank or trade references if required. It should be borne in mind by those inexperienced in building that balance sheets and financial reports of construction companies can be complex and to the uninitiated, misleading. Care should be taken in interpreting such financial information.

A Standard Form of Tendering Questionnaire (Private Edition) has been prepared by the NJCC for completion by contractors who wish to be included on tender lists. Further particulars must be supplied if this form is used, in order to provide contractors with the basic information concerning the project.

3.07 ANALYSIS AND RESPONSE BY THE CONTRACTOR

On receipt of an enquiry requesting a contractor to prequalify for a project, a basic record form must be completed. This form will summarise for management the basic data concerning the project. At this stage all of the information may not be available and there are advantages in recording the data on a form which will also be used when a firm enquiry to tender is received. This will allow the sequential completion of the form as information is available. The **'Preliminary/Tender Enquiry Form'** is shown on Page 14.

The task of completing the preselection questionnaire and providing supplementary data to support any request for prequalification often falls to the estimating department. As a consequence, management must ensure that the estimating department remains fully informed of the company's work load objectives, projects in hand and completed, and that it has access to the required financial and personnel data concerning the company that is sought from preselection questionnaires.

The final submission, prepared and presented by management, will take note of the recommendations of the estimating department on the following points:

- suitability of the project with regard to the company's future work load requirements;
- financial resources and commitments of the client, related to the value of the project;
- market conditions, including the general economic and political situation and outlook;
- construction problems;
- previous experience with similar types of project;
- previous experience with the client or consultant;
- adequacy of information provided;
- resources needed to carry out the project in accordance with the client's requirements;
- workload of the estimating department;
- seasonal risks;
- risks imposed in the conditions of contract;
- risks in the construction of the project;
- the number of contractors to be invited to tender.

3.08 ANALYSIS OF THE CONTRACTOR'S RESPONSE

Preliminary information

The preliminary information sent to the contractor will enable him to determine whether he wishes to tender for a project (this is considered in more detail in Section 4 'Decision to Tender'). The response from the contractor, if positive, will indicate to the client and his consultants whether the contractor has the necessary expertise, experience, management, resources (including financial) and knowledge of the location to adequately construct the works in the time required. Adequate and consistent appraisal of contractors submitting information is essential.

Interview

On larger projects it may well be necessary for the contractor to meet with the client and consultants for interview. This allows both parties to elaborate on their proposals and to discuss the project and method of working in more detail. It is important at this stage that the contractor's personnel who will be considered for the project should be present, as well as those who will prepare the tender. Also, the consultants who will be responsible directly for the project should attend, as well as senior representatives of the practices involved in the project. An early opportunity for key personalities associated with the future project to meet will be advantageous to all.

Once a contractor has indicated his willingness to submit a tender, he should only under exceptional circumstances, subsequently withdraw from submission of a tender. If a situation occurs which does cause the need for such a withdrawal, ie an unexpected increase in work load which commits the contractor's resources, then maximum notice should be given to the client or consultants of such withdrawal.

Selected list

From the written exchange of information, and any interviews felt necessary by the client and his consultants, a short list of contractors will be produced. This preselection procedure will produce a list of companies, any of which would be capable of constructing the works.

PRELIMINARY/TENDER ENQUIRY FORM

Ref. No.

A PROJECT PARTICULARS

Project description	Project address
Date enquiry received	Date tender to be submitted
Project type	Client
Architect	Engineer	Quantity Surveyor
Drawings available at	By arrangement with

B TENDER INFORMATION

Has the project been the subject of previous invitations to tender/preselection YES/NO	Type of tender Open Selective Negotiated Other
Proposed date of issue of tender documents	Proposed tender period	Acceptance period for tender
Estimated cost of project	Own work + Domestic sub-contractors + Labour-only sub-contractors + PC and provisional sums + Others = Total

14

C PROJECT DETAILS

Description of project	Contract period stated in documents	Own assessment of contract period	Is contract duration critical?
Special features	Date for commencement		Date for completion
Adequacy of information provided	Phasing requirements		

D CONTRACT DETAILS

Form of contract				Method of measurement		
Amendments to conditions of contract or special conditions			Insurances		Bonds	
Fluctuations recovery based on basic list	Clause		By contractor		By consultant	Fixed price
Fluctuations recovery based on formula	Clause	Type of formula	Deductions		Base month	
Period of interim certificates		Period for honouring certificates		Minimum amount of certificates		
% value of materials and goods to be certified		% of certificate value retained		Limit of retention withheld		
Period of measurement		Defects liability period		Liquidated damages		

E ESTIMATORS REPORT

continued overleaf

F DECISION

To submit preselection details	YES/NO
To submit a tender	YES/NO

Consultants should advise contractors who are selected at this preselection stage that they will be receiving tender documents on the advised date. The contractor can then reserve appropriate resources to process the tender when it arrives. It is important that these dates are adhered to and that the contractor is advised of any changes that occur. Contractors who are not selected at the preselection stage should also be advised promptly that they have not been successful.

ADDITIONAL READING

1. NATIONAL JOINT CONSULTATIVE COMMITTEE. Code of procedure for single stage selective tendering. 1977. RIBA
2. CHARTERED INSTITUTE OF BUILDING. Contractor selection — a guide to good practice. Estimating Information Service Paper No. 34. 1979
3. BARRETT, F. R. Cost value reconciliation. 1982. CIOB
4. NATIONAL JOINT CONSULTATIVE COMMITTEE. Standard Form of Tendering Questionnaire (Private Edition) 1983. RIBA.

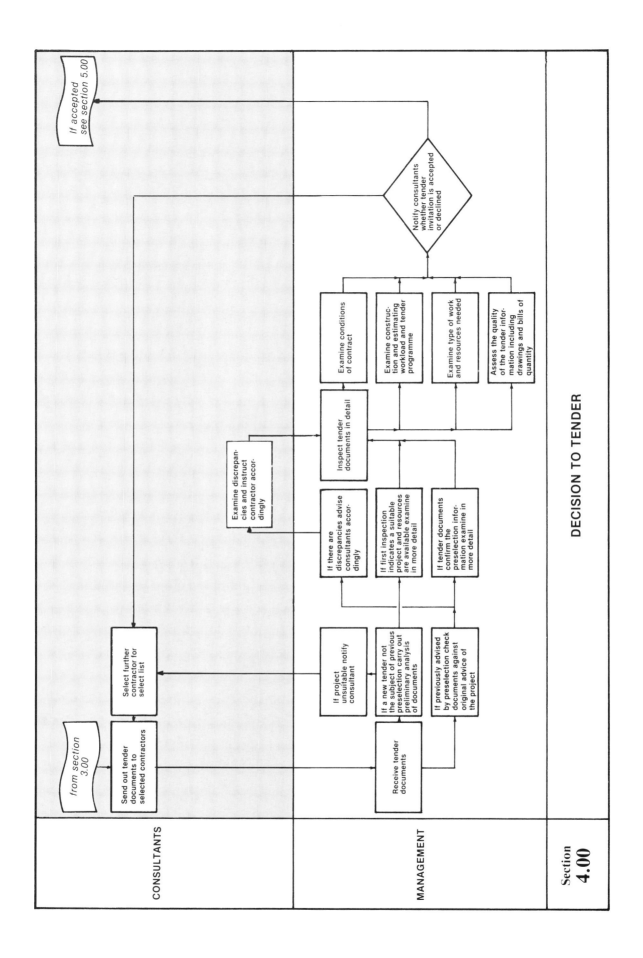

4.00 Decision to tender

4.01	INTRODUCTION	4.06	INFORMATION REQUIRED FOR THE PREPARATION OF AN ESTIMATE
4.02	RECEIPT OF TENDER DOCUMENTS		– *General* – *Site layout drawings*
4.03	PROJECTS FOR WHICH PRESELECTION HAS NOT OCCURRED		– *Other drawings* – *Detailed drawings etc* – *Technical reports* – *Special requirements of the client*
4.04	PROJECTS WHERE PRESELECTION HAS OCCURRED	4.07	CONDITIONS OF CONTRACT
4.05	INSPECTION OF THE TENDER DOCUMENTS	4.08	WORKLOAD AND TIMETABLE
		4.09	TYPE OF WORK AND RESOURCES NEEDED
		4.10	TENDER INFORMATION
		4.11	MANAGEMENT DECISION
			ADDITIONAL READING

4.01 INTRODUCTION

The decision to tender can occur at one of two stages:
- when preselection enquiries are initiated by a client or his consultants, the contractor will make a decision based upon an outline of the tender information available at that stage. This intention to submit a tender must be re-affirmed when the full invitation to tender and supporting documentation are received;
- when the preselection procedure has not been followed the contractor may find that tenders arrive without prior notice. In such instances, only one opportunity exists to appraise the project and make the necessary decision to tender for a project or not.

In the case of a project where preselection has occurred and details have already been sent to the contractor, a checking procedure is needed to confirm that the project conforms with the information already provided and that the contractor's position regarding tendering has not changed. It is essential that adequate time is allowed for this procedure by clients and consultants. In either situation, the contractor will follow the same procedures in analysing the information received.

4.02 RECEIPT OF TENDER DOCUMENTS

On receipt of tender documents a **'Preliminary/Tender Enquiry Form'** must be completed by the estimating department and an acknowledgement or receipt of the enquiry sent to the consultant. This form will provide management with a summary of the project and the tender documentation, and will be a significant document contributing to the decision concerning the submission of a tender.

Where an approximate cost of the project is provided this must be reconciled against any advice given at the time of preselection (see 3.06). If no approximate cost is given, an early assessment must be made by the estimator to determine the approximate cost of the project and the scope of the works.

4.03 PROJECTS FOR WHICH PRESELECTION HAS NOT OCCURRED

When the contractor receives tender documents for a project for which preselection has not occurred, there is increased urgency in deciding whether to tender. The documents may be unexpected and cause problems in an estimating department which already has a full work load. The project may be of particular interest to the contractor and so it is vital that an early appraisal is made of the potential of any new enquiry.

If the project cannot be accommodated because of work load in the estimating department, or the company's work load needs, then the consultant must be advised, at the earliest opportunity, that the contractor will not be submitting a tender. This will allow time for the selection of another contractor by the consultant if it is considered necessary to retain a full tender list.

If the initial assessment of the tender documents indicates that the project is of interest to the contractor and if estimating department work load commitments allow, then a full inspection of the tender documents will be necessary.

4.04 PROJECTS WHERE PRESELECTION HAS OCCURRED

The first operation will be to check that the tender documentation conforms with the information given at the time of preselection. If this initial inspection discloses discrepancies in the project information, programme, conditions of contract or other areas, the consultants should be advised accordingly. The contractor will then need to examine the tender documents in detail to establish whether the changes are of such significance to discourage him from tendering.

If the initial inspection confirms that the project conforms to the information given at the time of preselection, the contractor will now proceed with a detailed examination of the tender documents.

4.05 INSPECTION OF THE TENDER DOCUMENTS

The inspection of the tender documents must be made by the estimator who will be responsible for the production of the cost estimate and a checklist must be established of the documents received. In larger organisations, the documents may also be inspected by the chief estimator and other members of the contractor's organisation, including the programmer, quantity surveyor, buyer and contracts manager.

Clear lines of communication are needed to ensure that all viewpoints of those examining the documents are considered. The project estimator will be responsible for the co-ordination of these views.

Inspection of the tender documents must seek to achieve the following objectives:
- that the documents received check with the list of tender documents that should have been received, ie they are complete;
- that the documents and information are adequate for assessing costs;
- that sufficient time is available for production of the cost estimate and tender.

4.06 INFORMATION REQUIRED FOR THE PREPARATION OF AN ESTIMATE

General

For the preparation of a pre-tender construction programme, estimate and tender, all the information contained in Section 3 will be required, together with additional information. This may be supplied in various forms, including:

- drawings;
- specifications (including performance specifications);
- schedules;
- technical reports;
- programme work periods for major nominated sub-contractors;
- bills of quantity .

All documents must show the date of original issue and the date and nature of subsequent amendments.

Site layout drawings

Drawings must show the existing circumstances of the site and adjoining areas and should include:

- site boundaries;
- means of access, including approach roads and entrances, and any restrictions on access;
- contours;
- roadways and fences;
- wooded areas, water and other natural features;
- overhead, surface and underground lines and equipment of statutory undertakings and other owners;
- position of buildings on site to be demolished;
- position and height of buildings on site to be retained;
- position and height of structures adjoining the site;
- position and depth of existing foundations and groundworks;
- temporary structures;
- outline of new buildings and siteworks;
- working areas;
- restricted areas;
- location of strata investigations;
- 1/2500 Ordnance Survey block plan showing roads, natural features and site orientation, sufficient to permit the location of the site to be ascertained.

Other drawings

- general arrangements for the project, shown by plans, sections and elevations;
- works to adjoining structures;
- details of items requiring mechanical lifting, including size, weight, location and fixing needs;
- in the case of alteration work, surveys of the existing structure;
- any specifically designed structural temporary works;
- any special risks, construction methods proposed and sequence related to engineering considerations.

Detailed drawings, outline drawings and schedules

Detailed drawings will be required to give information sufficient to determine the location, nature and dimensions of work to:

- sub-structure;
- frame;
- upper floors;
- roof;
- canopies, balconies and projections;
- rooflights;
- staircases;
- external cladding;
- non-standard joinery;
- purpose-made doors and metal windows;
- structural walls;
- non-structural walls;
- suspended ceilings;
- drainage internal;
- drainage external;
- external works;
- builder's work in connection with services.

Outline drawings will be required to give information sufficient to determine the approximate location and nature of work to:

- waste, soil and overflow pipes;
- cold water services;
- hot water services;
- heating services;
- ventilation services;
- gas services;
- electrical services;
- other services.

Outline drawings or schedules will be required to work to:

- partitions;
- fittings;
- sanitary fittings;
- non-standard joinery.

Schedules are required of:

- standard joinery;
- standard windows, frames and ironmongery;
- standard doors, frames and ironmongery;
- miscellaneous ironmongery;
- external wall finishes;
- internal wall finishes;
- floor finishes;
- ceiling finishes;
- decorations.

Technical reports

- strata information, including water table;
- special site conditions such as fire risks, security risks or radiation hazards;
- other technical reports.

Special requirement of the client

- provision of drawings of services as installed, for maintenance purposes;
- training of client's staff in operating services or mechanical equipment.

4.07 CONDITIONS OF CONTRACT

Many standard forms of contract will be encountered in building. These include JCT'63, JCT'80, in both local authority and private versions, and with or without quantities alternatives, GC/Works/1 (lastest edition) used on central government projects, the Agreement for Minor Building Works used on smaller works, and many other more specialist forms used by larger public sector and private companies.

The particular contract conditions must be identified and any changes to standard forms of contract noted. Minor alterations can significantly affect the contractor's cash flow, the funding needs of the project, the risk allocation and the responsibilities of the respective parties to the contract. Such changes must be highlighted by the estimator and information passed to management on the extent of the inherent risk to the contractor of any changes made. Such advice will be of significance in any decision to tender.

4.08 WORK LOAD AND TIMETABLE

Two separate decisions must be made concerning workload. The chief estimator must make the decision concerning the workload of the estimating department, and be satisfied that the estimator who will be allocated to a project has the necessary expertise and knowledge needed for that particular project. The advantages of preplanning, which will be possible from advance warning by preselection, will be obvious.

The second decision must be made by management. This concerns the objectives and needs of the company for workload and also the availability of resources to construct the project. Management must be satisfied that the project meets the company's objectives regarding workload and that the company is not exposed to undue risk by commitment to excessive amounts of work with one client, sector of industry or location.

The tender timetable will be of significance to both the estimator and management. The project estimator will normally be responsible for managing the production of the estimate and it will be essential for consultants to allow sufficient time for the assimilation of the project information and the production of the estimate. Inadequate time may be sufficient reason for a contractor declining an invitation to tender.

4.09 TYPE OF WORK AND RESOURCES NEEDED

In making a decision to tender, the contractor will initially examine the type of work required in the proposed project. Many contractors concentrate on particular types of building, ie industrial, housing, etc., whilst others prefer to maintain a spread of work across various building types. Whatever the corporate strategy of a company, an objective decision must be made as to whether a new project fits into the company's work load objectives at that time. Such a decision must also consider the resources currently available to the company.

Resources include finance, staff and also labour, bearing in mind the particular skills and quality needed for a project, as well as the availability of materials and plant. The estimator contributes to this assessment with his knowledge of market conditions, trends in the costs of basic resources and in the identification of problem or risk areas.

The financial appraisal involves an assessment of the payment proposals in the conditions, any bonding requirements, the relative proportions of work to be carried out directly by the contractor, sublet to domestic sub-contractors or carried out by nominated sub-contractors. From these assessments, the funding requirements of the contract will be established and finance charges can be calculated.

Decisions in these matters must be taken by management. Careful definition of company objectives and policy will allow estimators to concentrate the resources of the estimating department on the most suitable projects for the company when the enquiry is received.

4.10 TENDER INFORMATION

The estimator is particularly concerned with the quality of the information provided at the enquiry stage. Inadequate or contradictory information will lead to a cautious approach by the estimator. Whilst the estimator should seek further information from the consultants in such cases, the estimator's approach may often be coloured when uncertainties exist and the project is not fully defined. Past experience may well indicate that the uncertainties continue into the construction phase, often with dire consequences for both the project and the contractor.

If bills of quantity are provided, they should be fully descriptive of the works in themselves, without the need for reference to other project information. Any deviations from the Standard Method of Measurement should be noted. Where bills of quantity are not provided, drawings together with a comprehensive specification will be required as described in 4.06.

The quality of design and evidence of co-ordination of services will be of significance to the estimator. The extent of design advancement, as reflected in the drawings provided, must be compared with the bills of quantity to ensure that the bills accurately reflect the fully designed work. Note should be taken of provisional and PC sums where design is not yet complete.

The estimator must obtain from the project information details which:

- relate to the contractor's intended method of working;
- impose restrictions;
- affect access;
- interrupt the regular flow of trades;
- affect the duration of the project;
- require specialist skills or materials;
- have a significant effect on the programme;
- are of major cost significance.

Such items affect costs and will guide the estimator during the production of the cost estimate. It is also essential that phasing requirements are clearly defined not only in the contract documents but also by illustration on drawings. Inadequate definition of such items will lead to problems concerning programming and definition of cost.

The estimator must co-ordinate the views of the various members of the contractor's team who examine the project information. He must provide management with a realistic appraisal of this information to enable a decision to be made on whether to tender for the project or not.

4.11 MANAGEMENT DECISION

Comments on the contract conditions, work load, type of work and resources needed and the tender documentation will all be taken into account by management in deciding whether a tender will be submitted.

If it is decided to accept the invitation, the procedure for the preparation of an estimate described in Sections 5, 6, 7, 8, 9, 10 and 11 of this Code should be commenced and the consultant informed that a tender will be submitted. If it is decided that the invitation to tender will be declined, a verbal response should be given and all documents should then be immediately returned to the consultant with the reason why a tender will not be submitted. This action, if taken, must be carried out as soon as possible after receipt of the tender documents.

In either case, the **'Preliminiary/Tender Enquiry Form'** should be filed in a register with the decision noted.

ADDITIONAL READING

1 MUDD, D. R. Administration of a tender. CIOB Estimating Information Service Paper No. 33. 1979
2 DAVIES, F. A. W. Preparation and settlement of competitive lump sum tenders for building work in *The practice of estimating*. 1981. The Chartered Institute of Building

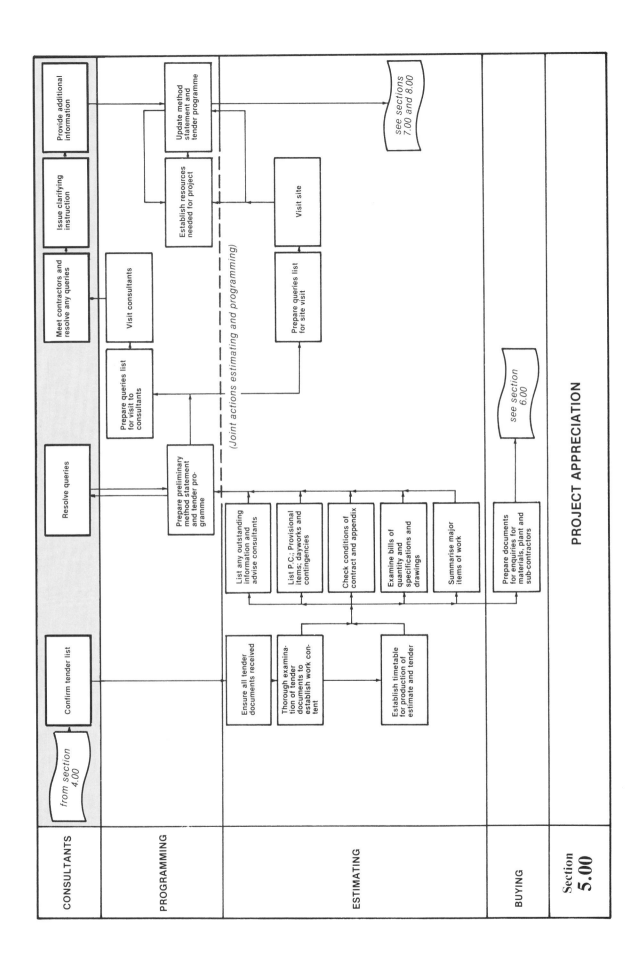

5.00 Project appreciation

5.01	MANAGEMENT OF THE ESTIMATE	5.05	PRIME COST SUMS, PROVISIONAL SUMS, DAYWORK AND CONTINGENCIES
5.02	ENSURE ALL TENDER DOCUMENTS HAVE BEEN RECEIVED		– *Prime cost sums* – *Provisional sums* – *Daywork* – *Contingencies*
5.03	TIMETABLE FOR PRODUCTION OF THE ESTIMATE AND TENDER	5.06	METHOD STATEMENT AND TENDER PROGRAMME
5.04	THOROUGH EXAMINATION OF THE TENDER DOCUMENTS	5.07	SITE VISIT
	– *Conditions of contract and appendix* – *Specification, bills of quantity and drawings* – *Outstanding information*	5.08	VISIT TO CONSULTANTS ADDITIONAL READING

5.01 MANAGEMENT OF THE ESTIMATE

Project appreciation commences in the 'Decision to Tender' phase but intensifies once management has confirmed that an estimate is to be prepared.

Considerable management skills and personal leadership will be needed by the estimator to motivate and co-ordinate the various staff associated with the preparation of the estimate and its subsequent conversion into a tender.

Co-ordination meetings may well be necessary with management and other departments within the contractor's organisation to establish key dates, decide on actions necessary and monitor progress during the production of the cost estimate. An example of a typical check list for such co-ordination is shown on Page 28.

5.02 ENSURE ALL TENDER DOCUMENTS HAVE BEEN RECEIVED

Once the decision has been taken to tender for a project, the estimator must ensure that all the tender documents have been received. A check must, therefore, be made to see that all drawings received are of the revision noted, and that all other documents are present.

A master register should be set up to list these documents, together with any supporting schedules or other data and the action taken with this information during the tender period recorded, eg which copies are sent to suppliers and sub-contractors. Any subsequent additional or revised documents should also be recorded in the register.

A record must be made in the **'Estimator's summary, analysis and report'** concerning key data and drawings available (see Page 118).

CO-ORDINATION MEETING CHECKLIST

Ref. No.

A | PROJECT DESCRIPTION |
| CLIENT |
| CONTRACT NAME/REFERENCE |
| CONTRACT ADDRESS |
| CONTRACT PERIOD | TENDER DATE |

B | DATE OF CO-ORDINATION MEETING | HELD AT:- |
| PRESENT:- |

C

LATEST DATES FOR:-	DATE	ACTION	LATEST DATES FOR:-	DATE	ACTION
Despatch of enquiries			Complete cost estimate		
Receipt of quotations			Finalise programme		
Visit to consultants			Further co-ordination meetings		
Visit to site			Adjudication meeting		
Finalise method statement					

D

SUPPORTING DOCUMENTS		INCLUDED
Schedule of trades to be sub-contracted and relevant documents		
Major materials and plant items		
Preliminary/Tender Enquiry Form		
Special reports or information		

E SPECIAL FEATURES

5.03 TIMETABLE FOR THE PRODUCTION OF THE ESTIMATE AND TENDER

The estimator must, as the manager responsible for the production of the estimate, ensure that a timetable is established which highlights the key dates in the production of the estimate and tender. It will be an essential document for all those associated with the tendering function. The following dates must be established without ambiguity:

- latest date for despatch of enquiries for materials, plant and sub-contracted items;
- latest date for the receipt of quotations;
- visit to the consultants;
- visit to the site and the locality;
- finalisation of the method statement;
- completion of pricing and measured rates;
- finalisation of the tender programme;
- intermediate co-ordination meetings within the contractor's organisation;
- adjudication meetings;
- submission of the tender.

All personnel associated with the tender must confirm that they are able to provide the necessary data in the format required, in accordance with the agreed timetable.

5.04 THOROUGH EXAMINATION OF THE TENDER DOCUMENTS

The conditions of contract, bills of quantity, general arrangement drawings, specification and any supporting documents, such as soil reports, site industrial relations policy, or specialist information must be examined in detail by the estimator. The estimator has a responsibility to ensure that other members of the contractor's estimating team with specialist knowledge or interests should be provided with copies of any relevant project information.

At this stage, the tender documents must also be examined by other personnel in the contractor's organisation, including those associated with purchasing, programming, temporary works, plant and the construction phase. The estimator should maintain records of those personnel who receive copies of the tender documents. Any discrepancies or divergencies between any of the tender documents must be noted by the estimator for future reference to the appropriate consultant for clarification.

Conditions of contract and appendix

Any unusual features in the conditions of contract and appendix will have been identified prior to the decision to tender and noted on the '**Preliminary/Tender Enquiry Form**'. At this stage the estimator must consider in more depth the consequences of such conditions and note on the '**Estimator's summary, analysis and report**' his recommendations for dealing with the situation.

Any further particulars received during the tender period concerning the conditions of contract and appendix details, must also be entered on the register and the '**Estimator's summary, analysis and report**' and the estimator must ensure that appropriate action is taken regarding any cost implications of revised or additional information received.

The examination of conditions of contract and appendix must reveal whether:

- non-standard forms of contract are used;
- conditions of standard forms are amended;
- non-standard payment or retention provisions are included and any possible affects that such alterations may cause regarding the contractor's liability for payments to sub-contractors;
- bid or performance bonds are required;
- insurance requirements are met by the contractor's standard policies;
- nomination procedures will be followed without amendment;
- liquidated damages requirements are acceptable and whether any allowance should be made in the tender;
- amendments to other project information have been reflected in the condition of contract, ie items not measured in the bills of quantity in accordance with the SMM.

Specification, bills of quantity and drawings

The estimator must have a thorough understanding of the specification. This document must be examined in detail and compared with the items measured in the bills of quantity and tender drawings, to identify any divergencies and discrepancies.

If clauses from the National Building Specification (NBS) are used, the estimator must ensure that they are unaltered and that there is consistency in use.

The specification must be clear and unambiguous and the estimator must prepare a queries list of any ambiguous items for discussion with the appropriate consultant.

A schedule must be prepared for use in obtaining quotations for various materials and trades to be sub-contracted. In order to prepare this the estimator must first of all group together relevant items from the specification and bills of quantity into respective trades and if necessary, individual suppliers. It may also be necessary to obtain further information from drawings, for example, lengths of timber may need to be specified. The estimator should consider any additional quotations that will be required to enable him to have flexibility when pricing. This could include extra over rates for wrot timbers, grading, treatment or small delivery surcharges. Such information will be of benefit when pricing the bills in detail at a later stage. (A sample sheet for abstracting specification and bill of quantity items for materials and sub-contractors enquiries is shown on Page 31. **'Abstract for materials enquiries'** and Page 32, **'Abstract for sub-contractor enquiries'**.

The bills of quantity must be carefully analysed and any divergencies from the SMM noted and drawn to the attention of the consultants.

Attention must be paid to any small quantities of similar materials which are noted within several sections of the bills of quantity. When aggregated together, such items may form a significant total, but their use on site may require a series of small deliveries at premium rates. Notes should be prepared of such requirements for use when enquiries are sent out and when unit rates are being prepared.

The information required on tender drawings has already been described in Section 4 and will be a significant factor in the decision to tender. At this stage, the estimator will systematically analyse the data shown on the various tender drawings and schedules and build up a picture of the project's needs. This analysis must highlight the cost significant items in the project and demonstrate areas of caution or risk. Further information may be needed when the consultants and site are visited. This information must be recorded for adjudication purposes in the **'Estimator's summary, analysis and report'**.

ABSTRACT FOR MATERIAL ENQUIRIES

PROJECT

Ref. No.
Page of

| SPECIFICATION ||| B.Q. ||| DESCRIPTION | APPROX QTY | UNIT | SPECIFICATION ||| B.Q. ||| DESCRIPTION | APPROX QTY | UNIT |
Bill	Page	Item	Bill	Page	Item				Bill	Page	Item	Bill	Page	Item			

ABSTRACT FOR SUB-CONTRACTOR ENQUIRIES

PROJECT

Ref. No.
Page of

TRADE	TRADE	TRADE	TRADE	TRADE	TRADE

Bill	Page	Item	Bill	Page	Item	Bill	Page	Item	Bill	Page	Item	Bill	Page	Item	Bill	Page	Item

Bill	Page	Item	Bill	Page	Item	Bill	Page	Item	Bill	Page	Item	Bill	Page	Item	Bill	Page	Item

The estimator should look for various factors at this stage which may influence his approach to pricing such as:

- standard and completeness of the drawn information;
- tolerances required;
- clarity of the specification requirements and the quality required;
- buildability;
- whether load bearing and non-load bearing areas can be identified;
- the extent of use of standard details indicating previous construction experience;
- evidence of design co-ordination of services and structural needs;
- the amount of information concerning ground conditions and foundations;
- problem areas and restraints on construction in the design.

From this appreciation of the drawn information, the specification and measured items in the bills of quantity, the estimator will begin to understand how the job is to be built and the quantity and quality of resources needed. It is to be regretted that whilst designers often take several months or even years to arrive at this stage on a project, the estimator and his team will be required to assimilate all the information produced, decide on how the project will be built and estimate its cost of construction, often within a tender period of a few weeks.

Outstanding information

The thorough examination of the tender documents by the estimator will highlight queries concerning the project and the method of construction. It is essential that the estimator lists these and collates together queries that are raised by other members of the contractor's team who have examined the documents. It is important that all queries concerning the tender information are channelled to the estimator for resolution. Ease of communication and good management practice dictate that only the estimator should deal with the consultants at this stage.

Queries will be resolved by meetings and discussions between the estimator and other members of the contractor's team and the consultants or the site visit. The estimator must ensure that all queries raised with the consultant are made in writing and that the urgency of the information needed is stressed. Any significant decisions communicated to one contractor by the consultant, must be communicated to all tendering contractors.

5.05 PRIME COST SUMS, PROVISIONAL SUMS, DAYWORK AND CONTINGENCIES

Very often, a large proportion of a project is contained within Prime Cost sums, provisional sums, daywork and contingencies. It is essential that an early assessment is made of these sums in relation to the total value of the project. A preliminary record will already have been established on the **'Preliminary/Tender Enquiry Form'**, and more analysis of these sums will now be required. A **'Schedule of PC and Provisional Sums'** as shown on Page 34 will assist.

Prime Cost sums (PC sums)

PC sums may be included in bills of quantity for work to be carried out by nominated sub-contractors or materials to be provided by nominated suppliers. Alternatively, the contractor may be invited to tender for the provision of items covered by PC sums. The estimator must check that the items of work covered by a PC sum are clearly stated and that the work can be identified within the project.

SCHEDULE OF PC AND PROVISIONAL SUMS, DAYWORK AND CONTINGENCIES

PROJECT		Ref.No.

B.Q. PAGE ITEM NO.	PC AND PROVISIONAL SUMS, DAYWORK AND CONTINGENCIES	DESCRIPTION	Discount £	Nom. sub-con £	Nom. Supplier £	Net PC Sums £	General attend. £	Special attend. £
	TOTALS EXCLUDING DISCOUNTS	£						
		ADJUSTMENTS ADD/OMIT						
	GRAND TOTAL EXCLUDING DISCOUNTS	£						

COMMENTS

Other items to be checked include:

- whether the name of the nominated sub-contractor or supplier is known;
- that the nominated sub-contractors and suppliers will be completing design indemnity warranties and that the contractor assumes no design reponsibility for such works;
- the degree of advancement of such design work and co-ordination of such work on to other consultants' drawings;
- that attendance (both general and special) are adequately identified and measured in accordance with the requirements of the SMM;
- that adequate provision has been made for contractor's profit addition and that discounts are to be allowed;
- that the nominated sub-contractors and suppliers can conform with the requirements of the contractor's tender programme;
- that builder's work requirements are clearly defined and measured in the bills of quantity.

Often such information will not be available and the contractor will be required to make assumptions regarding the programming of the works of nominated sub-contractors and the delivery of materials provided by nominated suppliers. In such situations, the contractor's tender programme must clearly state the basis on which time has been allowed for such work.

Many queries can arise over PC sums, including defining the scope and extent of the work, attendance items, and the pricing of builder's work associated with nominated sub-contractors. (See Section 9, Attendances).

Provisional sums

A provisional sum will be included in bills of quantity for an item of work needed, but not fully defined at the time of tender. This sum will eventually become the subject of an architect's instruction during the construction phase, and the items will then be incorporated into the measured work of the contract, subject to the appropriate contract rules for measurement and valuation.

The estimator must recognise like items of work in the bills of quantity and ensure that items, which may increase as a result of the expenditure of a provisional sum, are accurately priced.

Daywork

Sums are included for daywork to allow for expenditure on unforeseen works, which are either more appropriately paid on a 'time and material' basis, or where items in the existing bills of quantity are not appropriate for the varied work.

The estimator must ensure that daywork is defined in accordance with *The definition of prime cost of daywork carried out under a building contract*. Once the estimator establishes that the definition of prime cost is satisfactory, he must ascertain the percentage addition to be added to the rate. Percentage additions must be calculated for labour, materials and plant, and great care will be needed to establish the correct additions, bearing in mind the changes that may be made to the basic data contained in the standard schedules used for daywork.

(The pricing of daywork is more fully described in Section 9).

Contingencies

Contingencies are a sum of money contained in the bills of quantity against some unforeseen expenditure on behalf of the client. A large sum in proportion to the total project cost might indicate that the project is not fully thought out and designed, that change might occur, or difficult construction, as yet undefined, may be encountered.

5.06 METHOD STATEMENT AND TENDER PROGRAMME

It is essential that an early meeting is held between the estimator and those responsible for the programming and construction of the project, to set out initial thoughts on the method of construction. In a smaller organisation, all of these roles may be undertaken solely by the estimator.

The following points must be considered when deciding on the method of construction:

- site location and access;
- degree of repetition;
- shape of the building;
- extent that the building works occupy the site and, therefore, the available space for storage, hutting, movement on the site etc;
- adjacent buildings, structures etc;
- company's experience of the type of building required and the location;
- the availability of labour;
- availability of materials;
- extent of specialised work and its relationship to the general construction;
- amount of work to be sub-contracted;
- amount that design indicates construction method, ie restraints, formwork striking times, special sequence of construction;
- plant requirements.

Decisions concerning the resources to be used on the project will take into account such factors as:

- location and availability of labour and management within the company;
- cost of recruiting additional labour, its availability, quality and quantity;
- amount of work to be sub-contracted;
- plant available within the company;
- availability of plant outside the company;
- availability of materials, including long delivery items;
- current and future projects in the area which may affect the supply of basic resources;
- quality of workmanship required;
- special requirements of the project, ie special plant or skills needed;
- overlap of operations needed to meet programme requirements;
- materials handling on site, storage, distribution and waste, time span of the project and seasonal influences on method of construction;
- quality and complexity of the work.

Alternative methods of operating, site layout and sequences of work should be evaluated at this stage and decisions made on the intended method of construction. Confirmation of this decision may well have to wait for the visit to the consultants and/or the site visit.

The tender programme is a vital document for the contractor. If the contract is secured, this programme represents the contractor's intentions at the time of tender and upon which the pricing of the works was based. Many standard forms of contract require the contractor to submit a copy of his programme at an early stage of the contract and the programme submitted may well be the tender programme. As it reflects the contractor's intentions, based upon the tender information provided, it is appropriate that it should be the document used.

The contractor will not always be required to submit a programme with his tender and may only be required to confirm by signing the form of tender that he will execute the works in accordance with the commencement and completion dates stated. However, good estimating practice dictates that the contractor must prepare a tender programme to:

- verify that the dates given for construction are possible;
- maximise the efficiency of his resources;
- price the time related elements contained in the estimate;
- establish the method and sequence of working.

The tender programme usually develops in two stages. A preliminary programme will be drawn up at an early stage of the project as soon as the project information has been assimilated. This must take into account:

- contract commencement and completion dates;
- client requirements for phasing or stage completion;
- other stated requirements or sequence of work, ie delivery of client's equipment;
- work to be carried out by directly employed labour;
- work to be carried out by domestic sub-contractors;
- work to be carried out by nominated sub-contractors;
- key items to be supplied by nominated suppliers.

The preliminary programme will thus identify the parameters of the project and the main resources needed. It will establish key dates for major portions of the work and provide basic information which can be used when obtaining quotations for materials, plant and work to be sub-contracted. It also provides essential information concerning the time available for works which are the subject of PC sums and early advice can be given to the consultants of any likely restraints or problems that are foreseen.

A clear basis now exists on which construction method and sequence can be developed, following the clarification of any queries by the consultants and the site visit. When such queries and constraints have been resolved, the tender programme can be produced from a preliminary programme, taking into account the agreed construction method and sequence. The method statement can also be finalised.

The method statement will outline the sequence and methods of construction upon which the estimate is based. It should indicate how it is intended to deal with the major elements of work and it should highlight areas where new or difficult methods are necessary or intended. It should be supported with details of cost data, gang sizes, plant requirement and supervision requirements.

The primary purposes of the method statement are:

- to establish the principles on which the estimate is based;
- to acquaint construction personnel of the resource limits which have been allowed in the estimate and to describe the method of working envisaged at tender stage.

SITE VISIT REPORT

Ref. No.   Page   of

HEADINGS		REMARKS
CONTRACT DETAILS		
Site address		
General description		
EXISTING SITE DETAILS		
Buildings		
Services:-	(a) Telephone	
	(b) Gas	
	(c) Water	
	(d) Electric	
Obstructions & restraints		
Fences or hoardings		
Trees, streams etc.		
Access		
Location in relation to:-	(a) Road	
	(b) Rail	
	(c) Public transport	
GROUND CONDITIONS		
Topography		
Borehole details		
Sub-strata:-	(a) Type of soil	
	(b) Stability of soil	
	(c) Water table	
	(d) Source of water	
	(e) Pumping requirements	
	(f) Disposal of water	
Disposal of soil		

HEADINGS	REMARKS
SERVICES (Nearest available)	
(a) Telephone + authority	
(b) Gas + authority	
(c) Water + authority	
(d) Electric + authority	
Drains	
Foul	
Surface water	
OTHER SERVICES and AUTHORITIES	
Local and statutory	
Local tip and charges	
Police (and police regulations)	
Garage	
Hospital	
Cafés/Shops/Amenities	
TEMPORARY WORKS and ACCOMMODATION	
Offices	
Cl. of Works	
Stores	
Welfare	
Toilets	
Fences & hoardings	
Compound	
Roads & hard areas	
Signboards	
Protection of adjacent buildings	
Demolition of buildings or obstructions	

HEADINGS	REMARKS
ANY OTHER DETAILS	
Mileage from office	
Other work in area	
Local sub-contractors	
Local suppliers	
Local plant	
General weather conditions:-	
(a) Rain — high/low/average	
(b) Winds	
(c) Exposure to frost	
Type of adjacent buildings:-	
(a) Industrial	
(b) Residential	
(c) Schools	
(d) Hospitals	
(e) Churches	
(f) Shops/offices	
NAMES OF PERSONS VISITING SITE	
(a) Estimator	
(b) Contracts manager	
(c) Programmer	
SKETCHES/PHOTOGRAPHS ATTACHED	YES/NO
SIGNATURE(S)	
Date	

The completed tender programme must clearly identify the contractor's intentions concerning the construction of the work. It must be used to check and balance the resources needed on the project, against those already contained in the estimate. It is an important document to be considered at the adjudication stage.

When design is incomplete at the tender stage, the contractor will find it beneficial to incorporate in the tender programme a series of key dates by which design information or instructions regarding PC sums must be given. It may well be that the contractor's estimate is based on early ordering of materials to obtain price advantage and/or long delivery periods of materials. It will be essential that design information is made available to reflect these intentions and secure the prices contained in the contractor's estimate.

The programming technique used will reflect the complexity of the contract and comment on programming procedures and techniques is outside the scope of this Code.

5.07 SITE VISIT

The visit to the site must be made once preliminary assessment of the project has been carried out and a provisional method and sequence of construction established. The estimator may be accompanied on such a visit by other members of the contractor's team.

As well as visiting the site itself, the opportunity should be taken to examine the general locality and to establish the extent of other building works in the area. Visits should also be made to local labour agencies and suppliers in the area. Any opportunity to visit excavations near to the site must not be missed and in this connection, the local authorities may be able to give advice on local conditions and of any excavations which may be visible adjacent or near to the site.

A comprehensive report of the site visit must be prepared and a sample form is shown on Pages 38, 39 and 40 **(Site visit report)**. Site photographs can often be a useful way of recording information for discussion and record purposes.

Points to be noted when making a site visit include:

- position of the site in relation to road and rail and other public transport facilities;
- names and addresses of local and statutory authorities;
- topographical details of the site, including note of trees and site clearance required;
- any demolition work or temporary work needed to adjacent buildings;
- access points to the site and any restraints on layouts that have been considered;
- ground conditions, any evidence of surface water or excavations indicating ground conditions and water table;
- facilities in the area for the disposal of spoil;
- existing services, water, sewers, electricity, overhead cables etc;
- any security problems, need for hoardings etc;
- labour situation in the area;
- availability of materials;
- weather conditions, ie high rainfall, winds etc;
- temporary roads;
- location of nearest garages, hospital, police and cafés;
- nature and use of adjacent buildings, ie industrial, residential;
- police regulations;
- local sub-contractors;
- restraints imposed by adjacent buildings and services, ie space available for tower cranes, overhang, etc;

- other work currently in the area, or shortly to start;
- availability of space for site offices, canteen, stores, toilets and storage;
- the effect any client requirements may have upon access, storage, movement or accommodation;
- special difficulties.

5.08 VISIT TO CONSULTANTS

The estimator must visit the consultants. Visits will normally be made to the architect, but it may also be advantageous to visit the consulting engineer, services engineer and quantity surveyor, in order to meet the personalities who will subsequently be involved with the project.

Detailed drawings, reports of site investigations and any other available information must be inspected and notes and sketches made of all matters affecting either construction method or the likely cost of work.

A critical assessment must be made of the degree of advancement and quality of the design. A well developed and well documented design may be indicative of a smooth running and possibly profitable project. A design which is obviously ill-conceived and incomplete may cause delays during construction and allowance should be considered for this in the tender. An adverse report may lead to the reconsideration of the decision to submit a tender.

The estimator, at this stage, must obtain information which was not included with the tender documents. If such details now exist, the estimator must request copies of the drawings concerned. If the consultant is not prepared or able to issue such details, then the estimator must compile detailed notes and take dimensions from the drawings in the consultant's office. Confirmation must be obtained that the information can be regarded as firm and taken into account in preparing the estimate and that all tenderers have been advised of this additional information.

Visits to consultants must result in clarification of outstanding queries. Any information that is needed, for example, to finalise scaffolding needs, clarify foundation type and depth or method of excavation, must be obtained from the consultants.

Investigations at the consultant's office should include the following:

- an examination of layouts, details and working drawings available;
- an assessment of the degree of complexity of the work;
- clarification of any phasing requirements of the project and the extent of work involved in each phase;
- determining the site boundaries and names of adjoining owners;
- resolving queries arising from the project information received;
- discussing location of spoil heaps, offices, services etc;
- discovering whether the consultant has any doubts about whether or not the project will go ahead;
- assessing how prepared the consultant is to proceed with the scheme;
- discovering whether the consultant has a particular construction method in mind, or whether any constraints on construction method exist;
- assessing whether the client is fully aware of his requirements;
  (Is the client indecisive? Are variations likely?)
- obtaining further particulars on PC sums, including names of intended subcontractors and suppliers;
- confirming that statutory permissions and approvals have been obtained;
- discovering how long the consultant has been working on the project;
- enquiring how the contract period was established;
- enquiring about the preferred occupation date, if the contractor is required to state his own construction period;

- if the project is a large one, enquiring whether the consultant has already completed a project of such magnitude;
- assessing whether the consultant's attitude and views suggest an understanding and experienced approach to the project.

Analysis of the replies to the enquiries will indicate the state of design and advancement of detailing. The observations and reports that are made following such visits by the estimator should be objective and concise. Difficulties and restrictions should be clearly stated, leaving the final appraisal and comment until the adjudication stage.

ADDITIONAL READING

1 CHARTERED INSTITUTE OF BUILDING. Programmes in construction. 1981
2 BROCKFIELD, W. F. Planning in relation to estimating in *The practice of estimating*. 1981. CIOB
3 SPRINGETT, D.E. The estimating team — some aspects of its work, ibid
4 FARROW, J.J. Tendering — an applied science. CIOB Occasional Paper No.1. 1976.

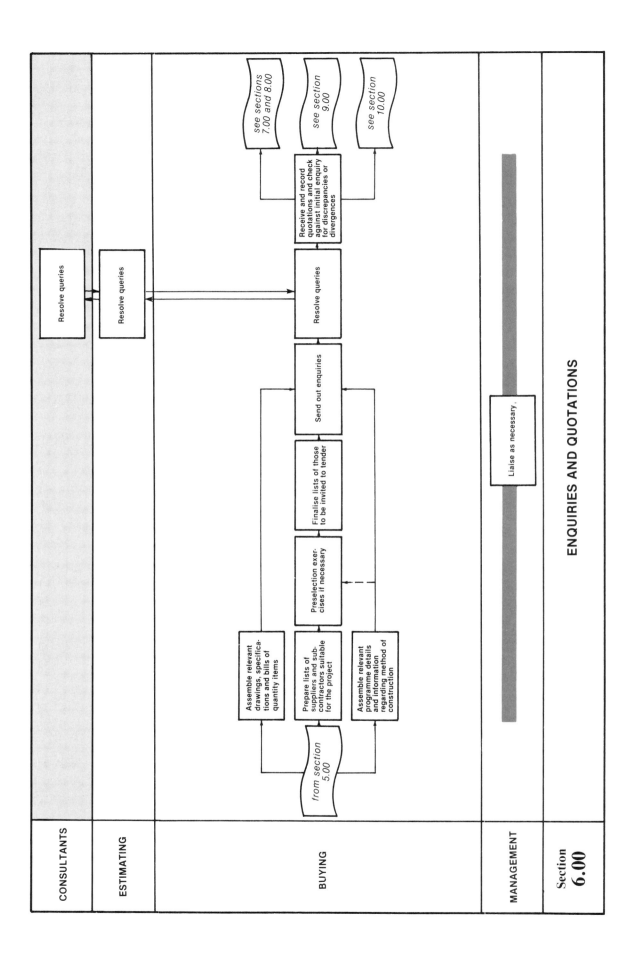

6.00 Enquiries and quotations

6.01	**PREPARATION OF DOCUMENTS FOR ENQUIRY PURPOSES**		– Hiring existing company owned plant
			– Hiring from external sources
	– *Lists and preselection*		– *Domestic sub-contractors*
	– *Programme and method of construction*		– *Labour only sub-contractors*
6.02	**ENQUIRIES**	6.03	**QUOTATION ANALYSIS**
			– *General*
	– *General*		– *Materials*
	– *Materials*		– *Plant*
	– *Plant*		– *Domestic sub-contractors*
	– Purchasing plant for the contract		**ADDITIONAL READING**

6.01 PREPARATION OF DOCUMENTS FOR ENQUIRY PURPOSES

The contractor's success in obtaining a contract can depend upon the quality of the quotations received for materials, plant and items to be sub-contracted. It is essential to obtain realistically competitive prices at the time of preparing the estimate.

The responsibility for carrying out this important function will vary from company to company. In some organisations the estimator will prepare the information for enquiry purposes, select the organisations to receive enquiries and reconcile the quotations received. In others, some or all of this will be done by the buying department, who may provide the estimator with a selection of fully reconciled quotations at the end of the enquiry stage for his final consideration. The accompanying chart shows this stage on a functional basis as buying. Standard procedures should be established, setting out the responsibilities of the person who is to carry out the enquiry function and any subsequent negotiations that might arise. The professional manner in which enquiries are sent to suppliers and sub-contractors can have considerable effect on the response and quality of quotations submitted.

Lists of items for which quotations are required will have been established following the detailed examination of the contract documents. If only drawings and specifications are provided it may be necessary for the contractor to produce additional information to assist with the enquiries. Every opportunity should be taken to clarify the contractor's requirements, bearing in mind the limited time usually available for obtaining quotations.

The contractor must ensure that comprehensive records are maintained of the various elements of project information sent to suppliers and sub-contractors. These records must list the drawings sent, the relevant contract and specification clauses, project preambles and the pages of bills of quantity.

The use of standardised documentation and procedures will assist in making this stage in preparation of the estimate routine, and allow an interchange of personnel at any stage.

Lists and preselection

Most contractors maintain comprehensive records concerning suppliers, and subcontractors. These records must include:

- a history of past performance;
- extent of geographical operation;
- size and type of contract on which previously used;
- information concerning contacts;
- address and telephone numbers.

When operating in a new area, a contractor will need information concerning the local suppliers and sub-contractors. In this case performance should be verified from other external sources and any remaining information established from the supplier and sub-contractor concerned.

A questionnaire may be used to establish the resources and abilities of sub-contractors concerning:

- area of operations;
- size and type of work;
- labour and supervision available;
- size and type of work previously carried out;
- references from trade, consultant and banking sources;
- insurances carried by the sub-contractor (if relevant);
- confirmation of holding of relevant sub-contractor's tax exception certificate (if relevant).

The questionnaire should attempt to establish the supplier or sub-contractor's financial capability to undertake the work in question and to supply the materials and plant required. It will be necessary to establish that all resources will be available to meet the requirements of the main contractor's programme.

Preselection will be more likely if dealing with unknown suppliers and sub-contractors. Bearing in mind the particular needs of the project, the contractor must ensure that the list of suppliers and sub-contractors who are invited to tender is comprehensive and that bids will be received. Preselection procedures must confirm that bids will be submitted and establish that the particular supplier or sub-contractor has the necessary resources and desire to tender for the project. The procedures described in Section 3 must be applied if sufficient time is available. In any event confirmation must be obtained from major suppliers and sub-contractors that they will be prepared to submit a bid, before enquiry documents are sent out.

Programme and method of construction

It is essential that suppliers and sub-contractors are advised of the programme requirements and any aspects of the method of construction that are relevant.

Programming information should include:

- the anticipated start date on site;
- the required completion date;
- key information of significance to the progress of the works;
- phasing.

Suppliers must be given an indication of delivery rates, together with any requirement for the approval of samples etc. Sub-contractors must be provided with details from the contractor's method statement which are relevant to their work.

6.02 ENQUIRIES

General

The objective of sending out enquiries is to produce technically accurate quotations and to set out clearly the terms and conditions on which quotations are being invited. It is recommended that limited numbers of suppliers and sub-contractors are invited to quote for each item or section of the work. This conforms with the principles of selective tendering and avoids the wasteful use of resources in requesting an excessive number of quotations.

The time allowed for the preparation of a tender is invariably short. The contractor has a limited period of time to distribute the relevant enquiries to a sufficient number of suppliers and sub-contractors to ensure an adequate response and yet at the same time limit the number of enquiries to a reasonable number of participants.

Clarity in the enquiry and its supporting documents is essential and every effort must be made to assist the supplier and sub-contractor concerned to avoid:

- the submission of incomplete quotations;
- failure to meet the required timetable for return of quotations;
- counter offers of alternatives which require further checking.

Each enquiry should fully describe the work to be undertaken or items to be supplied without ambiguity.

Materials

Enquiries to suppliers of materials should state:

- title and location of the work, and site address;
- specification, class and quality of the material;
- quantity of the material;
- likely delivery programme, ie period during which supplies would be needed with daily or weekly requirements where known. Where small quantities are to be called off from a bulk order this should be clearly stated;
- means of access, highlighting any limitations or delivery restraints;
- any traffic restrictions affecting delivery times;
- special delivery requirements, ie palletted, self-unloading etc;
- date by which the quotation is required;
- period for which the quotation is to remain open;
- whether fluctuating or firm price required, the basis for recovery of increased costs and the base date when a formula is used for calculation of fluctuations;
- discounts required;
- the person in the contractor's organisation responsible for queries.

The contractor has a responsibility to ensure that suppliers:

- make every effort to meet the specified 'date required'. If this is not posssible the contractor should ensure that he is informed promptly so that additional enquiries can be sent out in order to maintain a full enquiry list;

- clear queries as they arise. Quotations should not be submitted by suppliers with items marked 'more information required';

- submit the quotation on time with a clear statement where prices are 'to follow'. The latter should be dealt with without unnecessary delay.

The supplier must state in his quotation any special charges that may apply to his supplies. The contractor can incur costs for:

- pallets left on site;
- standing time for vehicles whilst unloading at the site;
- small quantities or abnormal loads;
- deliveries at abnormal times;
- unloading facilities to be provided by the contractor;
- special height, width or weight of deliveries which might affect the route or access to the site;
- any escort or special transport arrangements needed.

The contractor must record details of all enquiries sent to suppliers on a **'Materials quotation register'** (See Page 49) and await the receipt of the appropriate quotations.

Plant

The contractor's plant requirements will be established in the method statement and programme. They will establish the basic performance requirements of plant and in many cases will have identified specific plant items needed for the works. The duration for which the plant is needed on site will be established from the tender programme. The estimator must firstly compile a 'Schedule of plant requirements', listing the type, performance requirements and durations. This should be separated into:

- mechanical plant with operator;
- mechanical plant without operator;
- non-mechanical plant.

A note should be made on the schedule of any additional requirements associated with a particular item of plant, which must be provided by the contractor, ie a power supply for a tower crane, temporary access roads for erection purposes etc.

Further details are necessary for certain non-mechnical plant, eg scaffolding. The contractor must consider the intended method of working and programme requirements in the specification for plant. Turnround of equipment and striking time will dictate the amount of formwork and support and access equipment needed. A balance must be drawn between speed of operation and economy in establishing plant needs and all must be clearly reflected in the plant enquiry.

A list of plant suppliers must be established from companies who can meet the project's requirements. The options available for obtaining plant include:

- purchasing plant for the contract (in accordance with company policy);
- hiring existing company owned plant;
- hiring plant from external sources.

Purchasing plant for the contract

The decision to purchase plant for a particular contract will normally be taken by senior management. Such a decision requires a knowledge of plant engineering and will be made in accordance with the accounting policy of the company. Purchasing of plant is outside the scope of this Code but for guidance purposes only, the following general factors must be considered when plant is to be purchased for a project and sold on completion:

- purchase price less expected resale value after allowing for disposal costs;
- return required on capital invested;
- cost of finance;
- cost of maintaining the plant and associated overheads;
- stock levels of spares;

MATERIALS QUOTATIONS REGISTER

PROJECT									Ref. No. Page of
MATERIAL	QUANTITY	\multicolumn{6}{c}{QUOTATIONS RECEIVED}	QUOT'N USED	REMARKS
		1	2	3	4	5	6		

- the company's policy on depreciation;
- likely working life of the plant;
- cost of insurances and taxes, eg. road fund tax;
- any tax or depreciation allowances that are available;
- availability and cost of plant outside the company;
- accessability of the site in relation to company depot and servicing centres.

The manner in which the cost of such plant is charged subsequently to a contract, will depend upon the accounting policy of the company.

Hiring existing company owned plant

When plant is already owned by the company the contractor will be provided with hire rates at which plant will be charged to the site. The following list should be regarded, as guidance only, of the items which must be considered in building up hire rates for company owned plant:

- a capital sum based upon the purchase price and expected economic life; (This will vary according to the company's accounting policy.)
- an assessment of the costs of finance;
- the return required on capital invested;
- grants and financial assistance available when purchasing the plant;
- administration and depot costs;
- costs of insurances and road fund licences;
- maintenance time and costs and also cost of stocks needed for maintenance purposes.

Hiring from external sources

Where company owned plant is not available, enquiries must be sent to external suppliers for the plant required.

Enquiries for items of plant must either be sent specifying particular machines and equipment that are needed or specifying the performance required from the item of plant, eg tracked excavator with backacter, required to excavate trenches to a maximum depth of 3.00m, width of excavation 1.00m, 360° slewing required.

Enquiries should state:

- title and location of the worktand address of the site;
- specification of the plant or work to be done;
- anticipated periods of hire with start date on site and duration required;
- means of access, highlighting any restraints or limitations;
- any traffic restrictions affecting delivery times;
- anticipated working hours of the site;
- date the quotation is required;
- period the quotation is to remain open;
- whether fluctuating or firm price required, the basis for recovery of fluctuations and the base date when formulae are used for the recovery of fluctuations;
- discounts offered;
- the person in the contractor's organisation to be contacted regarding queries.

In addition to the basic hire charge per hour or week, the enquiry must seek to establish:

- cost of delivering and subsequently removal of plant from the site on completion of hire;
- cost of any operator, over and above the basic hire charge if provided by the hiring company. (If provided by the contractor the estimator must produce a

built-up rate for the operator's costs);
- whether the hire rates quoted include for servicing costs. If not the costs and timing of servicing must be established;
- any minimum hire periods applicable to the plant and the extent of any guaranteed time;
- cost of standing time and insurance costs if the plant is retained on site not working for any reason.

The contractor has a responsibility to ensure that suppliers:

- make every effort to meet the specified 'date required'. If this is not possible the contractor should ensure that he is informed promptly, so that additional enquiries can be sent out in order to maintain a full enquiry list;
- clear queries as they arise. Quotations should not be submitted with items marked 'more information required';
- submit the quotation on time with a clear statement where prices are 'to follow'. The latter should be dealt with without unecessary delay.

The contractor must record details of all enquiries to plant suppliers on a **'Plant quotation register'**, to await the receipt of suppliers' quotations (See Page 52).

Domestic sub-contractors

Sub-contractors will require the same details and information about the sub-contract works as the main contractor requires for his tender. Selection of sub-contractors will take into account skill, performance, integrity, responsibility and proven competence for work of similar size and character to the project under consideration and interest in tendering for the particular project. The principles of selective tendering relating to the main contract must be reflected in the number of sub-contract enquiries. The contractor's enquiry to sub-contractors must state:

- site, and location of the works;
- name of employer;
- names of the consultants;
- relevant contract and sub-contract details;
- any amendments to the main contract, appendix or sub-contract conditions;
- whether a fixed or fluctuating contract stating relevant details and rules;
- whether a schedule of labour and materials for a fluctuating contract if required;
- daywork rates as required;
- date quotations must be returned;
- general description of the works;
- particulars of access, available site plant, site industrial relations policy, storage facilities etc;
- where further details and drawings may be inspected;
- contract period, programme and method statement details, start date and duration for the sub-contract works. Any phasing requirements;
- any discounts required;
- two copies of the relevant extracts from the preliminaries and bills of quantity;
- drawings, schedules and reports where applicable;
- services or attendance to be provided by the main contractor (if any).

Opportunity must be provided for sub-contractors to see all the relevant drawings and details and every assistance given to ensure that the quotation received is the sub-contractor's best price and not inflated because of inadequate project information.

A simple procedure to avoid errors or misunderstanding is to provide photocopies of all relevant information as far as possible and to delete any irrelevant items not applicable to the enquiry in question. The objective of the enquiry is to provide all rele-

PLANT QUOTATIONS REGISTER

PROJECT			Ref. No.					
			Pageของ					

| PLANT ITEM | REF. | QUOTATIONS RECEIVED ||||||| QUOT'N USED | REMARKS |
| | | Basic hire rate | Operator cost | Fuels | Maintenance time & cost | Attendances required | Delivery & collection | | |

vant information and produce technically accurate quotations.

The contractor has a responsibility to ensure that sub-contractors:

- make every effort to meet the specified 'date required'. If this is not possible the contractor should ensure that he is informed promptly so that additional enquiries can be sent out in order to maintain a full enquiry list;
- clear queries as they arise. Quotations should not be submitted with items marked 'more information required';
- submit the quotation on time with a clear statement where prices are 'to follow'. The latter should be dealt with without unnecessary delay.

Labour only sub-contractors

The contractor may seek to obtain quotations from labour only sub-contractors for certain elements of the work. The decision to use labour only sub-contractors will take into account company policy, availability of labour in the area, any special requirements concerning site labour policy, and market conditions. In this case all items noted under 'Domestic sub-contractors' apply. The contractor should seek further information concerning the status of the labour only sub-contractor, particularly in regard to whether:

- the sub-contractor holds a current relevant sub-contractor's tax exemption certificate;
- the sub-contractor's insurances are adequate.

In addition the estimator must consider:

- any attendances to be provided by the contractor;
- responsibility for unloading, storage and distribution of materials;
- waste allowances;
- supervision to be provided by the contractor;
- frequency of payments required by the sub-contractor;
- retention sums relating to the sub-contracted work;
- any additional costs that may arise once the sub-contractor has completed the works and left the site;
- any additional risk.

The contractor must record all enquiries to sub-contractors on a **'Domestic sub-contractors' register'**, to await receipt of the sub-contractors' quotations (See Page 54).

6.03 QUOTATION ANALYSIS

General

The receipt of all quotations must be recorded methodically on the appropriate quotations register. Upon receipt of all quotations a detailed examination is necessary to verify that they have been made in accordance with the requirements of the enquiry. Even though precise information has been given to suppliers and sub-contractors, the contractor will frequently find that counter offers have been submitted. In all cases the contractor must ensure that all additions, extensions and collections are correct and that no pricing errors have occured.

DOMESTIC SUB-CONTRACTORS' REGISTER

PROJECT:

TRADE:

ENQUIRY LETTER USED:

Ref. No.:
Page ___ of ___

	NAME OF SUB-CONTRACTOR	Date of quotation	Sub-contractor's conditions	Quotation amount £	Discount %	Gross amount (abstracted from B.Q. & checked) £	Deduction for discount £	Net amount from B.Q. abstract £	Firm price allowance (if required and not included by s-c) %	Firm price allowance Amount £	TOTALS Gross £	TOTALS Net £
1												
2												
3												
4												
5												
6												

ESTIMATOR'S COMMENTS

Materials

Quotations from suppliers must be recorded on the **'Materials quotations register'**. Discrepancies and divergencies from the enquiry must also be recorded on the register for later evaluation, before selection of the quotation to be used in pricing unit rates. Quotations should be checked to ensure that:

- the materials comply with the specification;
- the materials will be available to meet the requirements of the construction programme;
- no special delivery conditions have been imposed by the supplier;
- the method of delivery complies with the contractor's requirements and intended method of unloading and handling on site;
- the conditions contained in the quotations do not amount to a counter offer being at variance with the terms and conditions of the enquiry;
- the quotation is valid for the required period;
- prices are given for small quantities and other items if required;
- discounts conform to the requirements of the enquiry;
- requirements concerning fixed or fluctuating prices are complied with.

The quotation should be examined to ensure completeness to identify any counter-offers of material which may not be acceptable or where items have been priced 'as price list'. Standard price lists, as for example often provided when pricing vitrified pipes and fittings, are not always simple, and care should be taken to arrive at the appropriate price when using such lists.

Plant

Quotations from plant suppliers must be recorded in the **'Plant quotations register'**. Discrepancies and divergencies from the enquiry must be identified and also recorded on the register. It must be borne in mind that where a performance specification was provided for the plant, quotations should be carefully checked to ensure that the plant offered meets the requirement defined at the enquiry stage.

Quotations must be scheduled to ensure that:

- the plant complies with the specification;
- it is available to meet the requirements of the construction programme;
- delivery and collection charges can be identified;
- where applicable, all operator costs are included and the operators will comply with the intended working hours of the site;
- any attendance or supplies to be provided by the contractor are clearly identified;
- maintenance responsibilities and charges are identified;
- maintenance liabilities are identified;
- the quotation is valid for the required period;
- the quotation conforms to the terms and conditions of the enquiry and does not represent a counter-offer;
- requirements concerning fixed or fluctuating prices are complied with.

Domestic sub-contractors (including labour only sub-contractors)

Quotations from sub-contractors must be recorded in the **'Domestic sub-contractors register'**. This procedure is similar to that described for the suppliers. Any divergences or discrepancies from the enquiry must be identified and highlighted for later evaluation at the time of selection of the quotation to be used in pricing unit rates. Quotations should be checked to ensure that:

- work described in the quotation complies with the specification and bills of quantities;
- all items have been priced;
- all items are priced correctly and in accordance with the unit of measurement billed;
- unit rates are consistent throughout the quotation;
- rates are realistic and comparable with those of competitors;
- the quotation is not a counter-offer and that the sub-contractor has accepted the terms and conditions of the enquiry;
- any attendance required from the contractor is clearly defined and acceptable;
- any ambiguities regarding responsibility for unloading, storage, protection or clearing up, etc, are resolved;
- discounts comply with the requirement of the enquiry;
- all supporting information such as schedules of rates, programming information, etc, are provided;
- requirements covering fixed or fluctuating prices are complied with.

ADDITIONAL READING

1 INSTITUTE OF BUILDING ESTIMATING SECTION. Sub-contractors — invitation to tender and adjudication of quotations. Southampton Group in *The practice of estimating*. 1981. CIOB
2 MUDD, D.R. Suppliers — invitation to tender for the supply of materials. ibid
3 TAYLOR, H. F. The link between estimator and buyer. ibid

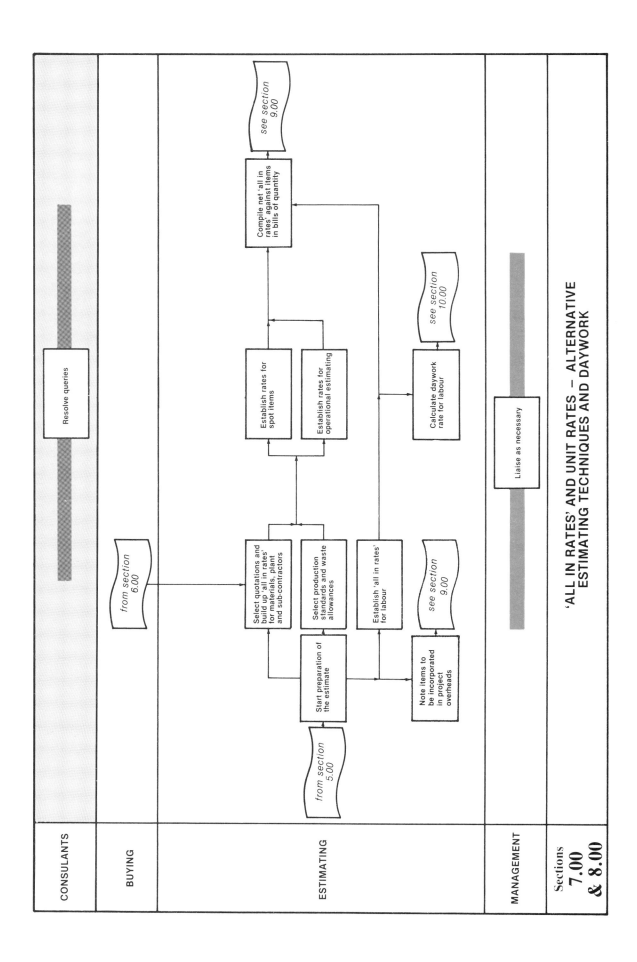

7.00 'All in rates' and unit rates

7.01 SUMMARY OF PROCEDURE FOR ESTABLISHING 'ALL IN RATES'

7.02 ESTIMATING 'ALL IN RATES' FOR LABOUR (STAGE 1)

- *Items for consideration in 'all in rates' for labour*
  - Labour costs normally contained in the 'all in rate'
  - Labour costs normally contained in project overheads

7.03 EXAMPLE OF THE CALCULATION OF AN 'ALL IN HOURLY RATE' FOR LABOUR

- *Summary of procedure for establishing 'all in rates' for labour*
- *Determination of hours worked*
  - Annual and public holidays
  - Calculation of hours worked
  - Inclement weather
- *Cost calculation*
  - Guaranteed minimum wages and emoluments (annual costs)
  - Additional emoluments
  - Overheads in employing labour
  - Summary
- *A standard form for calculation of an 'all in rate' for labour*

7.04 RECEIPT, ANALYSIS AND SELECTION OF QUOTATIONS

7.05 ESTABLISHMENT OF NET UNIT RATES (STAGE 2)

- *Principles*
- *Production standards and other considerations*
  - Labour element
  - Materials element
  - Plant element
  - Allocation of costs
  - Domestic sub-contractors element (including labour only sub-contractors)

ADDITIONAL READING

7.01 SUMMARY OF PROCEDURE FOR ESTABLISHING 'ALL IN RATES'

The preparation of an estimate falls into three distinct stages.

Stage 1: The establishment of 'all in rates' for the key items that will be incorporated. It includes:

- a rate per hour for the employment of *labour*. Different rates will be established for the different categories of labour that will be used on a project;
- an operating rate per hour (or per day, per week etc) for an item of *plant*. This may be plant supplied with or without operator and the rates may be established from the contractor's own data or from quotations received from plant hire organisations;
- a cost per unit of *material* delivered and unloaded at the site. This will involve comparison of the various quotations received for materials (see Section 6) and the selection of one of these for use in the estimate.

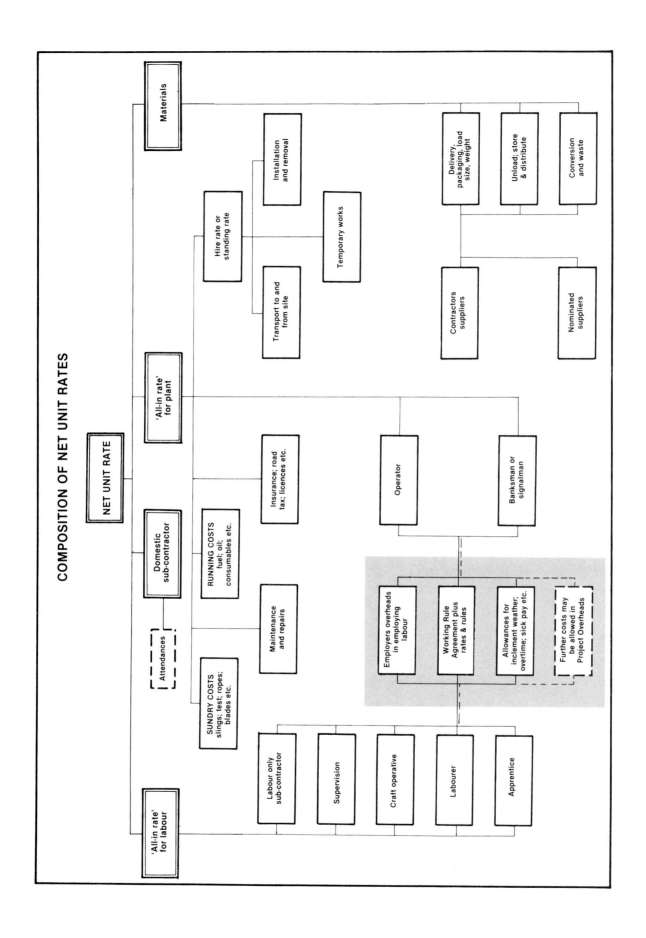

The cost of labour and some plant items can usually be established after the visits to the site and consultants. It is also necessary to complete the tender programme and method statement before finalising these prices. It will be necessary to await receipt of quotations in respect of materials, some plant items and sub-contracted work. (See Section 6). The preparation of 'all in rates' is shown diagramatically on the accompanying chart (See Page 60).

Stage 2: The selection of methods and *'production standards'* from the contractor's data bank or other sources. These standards are then used in conjunction with the 'all in rates' calculated in *Stage 1,* to calculate net unit rates which are set against the items in the bills of quantity. Alternatively, rates received by means of sub-contractor's quotations (see Section 6) may be used if the works are to be sub-contracted.

Stage 3: The calculation and addition of project overheads and the preparation of reports for consideration by management.

(Note: This Section of the Code covers *Stages 1* and *2. Stage 3* is considered in Section 9).

7.02 ESTIMATING 'ALL IN RATES' FOR LABOUR (STAGE 1)

Items for consideration in 'all in rates' for labour

Labour costs can arise in two areas. Costs associated with the Working Rule Agreement and certain overhead costs incurred by the employer.

Other costs, which will be variable and may be specific to a project, or some time related costs, will normally be contained in the project overheads.

Labour costs normally contained in the 'all in rate'

These can include:

Guaranteed minimum wages and emoluments paid to operatives

(a) The basic rate and guaranteed bonus, (or the contractor's bonus allowance).

Additional emoluments paid to operatives

(b) Inclement weather allowance;
(c) Non-productive overtime costs;
(d) Sick pay allowance — this will take into account a calculation for average sickness of operatives involving payments by the employer. Some companies operate private sickness schemes for operatives and the example does not take account of any special schemes. Where special sick pay schemes and benefits are operated, a decision must be taken as to whether allowances should be made in 'all in rate' or in project overheads;*
(e) Trade supervision;
(f) Payments due under Working Rule Agreement (extra payments).

---

*The sick pay allowance is based upon the Statutory Requirements set out in June 1983.

Overheads incurred by the employer in employing labour

(g) CITB training contributions;
(h) National Insurance contributions;
(i) Holiday credits – annual holidays;
             public holidays;
(j) Redundancy payments (statutory scheme);
(k) Employers' liability insurance.

Labour costs normally contained in project overheads

These can include: (see Section 9)

Guaranteed time

(l) Under the Working Rule Agreement, – 'Working Hours', the normal working hours are 39 hours throughout the year. (Monday-Thursday 8 hours per day, Friday 7 hours). As the hours used in the example are in excess of 39 hours, no allowance has been made for guaranteed time. Should the hours used in a calculation on any contract be less than the guaranteed time then the difference must be taken into account in project overheads under 'Guaranteed time'.

Pensions

(m) Although pensions are required under government schemes, many firms have 'contracted out' into private schemes. Therefore, any allowances to be made should be calculated in project overheads.

Daily travel allowance and fares

(n) Under the Working Rule Agreement, daily travelling is calculated on a sliding scale, depending on the distance in kilometres between the operative's home and his place of work. The total labour costs in the cost estimate should be used as a basis for calculation of the allowance in the project overheads for such expenses. (A detailed understanding of the Working Rule Agreement is essential in making such calculations). The provision of transport is normally priced in project overheads.

Lodging/subsistence allowance

(o) The costs involved for lodging and subsistence payments should be calculated using the total labour costs in the cost estimate, as mentioned previously. Allowance is made in the project overheads.

Abnormal overtime to meet time targets

(p) Allowances should be made in project overheads.

Attraction money

(q) Attraction money may be required to obtain labour in remote and difficult areas, where unpleasant conditions are encountered, or in areas where the volume of work causes shortages of local labour. Such sums must be carefully assessed and allowed in project overheads.

Special severance payments

(r) Special severance payments and/or terminal bonuses outside the legislative requirements concerning redundancy and termination may be considered necessary on a contract. Due allowance must be made in the project overheads

Other miscellaneous additions

(s) For example, computer payroll, labour turnover etc. These items can only be dealt with on an individual basis and should be allowed for in the project overheads.

It may not be possible to determine all of these factors with accuracy at an early stage of estimating, particularly as the volume of labour is not quantified. There are benefits in separating them from the calculation of 'all in rates' and for necessary allowances to be made in the later stages of the estimating process, in the project overheads.

It should also be noted that it is a matter of opinion and company preference where many of these items are priced. The distinction drawn between items set out in 'all in rates' and project overheads should not be regarded as mandatory. The important consideration is that due recognition must be paid to all items to establish the true costs involved and that adequate allowances are made in the estimate.

'All in rates' can be built up on a weekly basis, by considering an annual situation or the time period relating to a particular contract. In this Code the 'all in rate' is calculated on an annual basis.

7.03 EXAMPLE OF THE CALCULATION OF AN 'ALL IN HOURLY RATE' FOR LABOUR

Note: It is emphasised that the calculations in the following are examples only.

Summary of procedure for establishing 'all in rates' for labour

An 'all in hourly' rate should be built up in accordance with the principles established under 7.02.

Details vary according to:

- the actual trade of the operative (whether craft operative, labourer or mechanical plant operator);
- the firm;
- the area;
- the industrial and legal conditions in force at any time.

In this example the calculation is based on a one year employment period, and could apply to all projects for which tenders are to be submitted. It is for a craft operative.

Amendments must be made each time there is a variation in the cost of one of the factors included in the calculations, or when further factors are introduced. Alterations should be made as soon as variations are promulgated, though there may be a period of time before they come into effect.

The whole calculation should be revised regularly. There are four stages in the calculation:

1. Determine the number of working hours which an operative is expected to work during the one year period;
2. Calculate the cost per year for wages and the cost of each item used from the list contained in 7.02;
3. Summarise the individual costs obtained in (2);
4. Obtain a rate per hour by dividing the total costs in (3) by the number of hours in (1).

Alternatively, for a very large project, it may be desirable to make a special calculation based on the anticipated construction period.

Determination of hours worked

The number of hours worked during the calendar year, ie January to December, will depend upon the hours worked per week during the Summer and Winter periods, with adjustments for annual holidays and public holidays. The hours worked will vary between different companies and some variation in hours can also be expected between firms operating in the North and South of the country, due to the available amount of natural daylight hours in the winter period. Local customs and availability of labour will also affect the number of hours worked. Furthermore, a company may also agree to work hours suitable for a particular type of contract, due to special requirements of the employer or by special agreement with its employees.

The Working Rule Agreement, states that a 39 hour week should be worked throughout the year as follows:

— Monday to Thursday at 8 hours per day;
— Friday 7 hours.

For calculation purposes only, the hours used are typical of those worked on many sites. The summer period commences when British Summer time begins (Sunday 27 March 1983) and ends (Sunday 23 October 1983).

Annual and public holidays

The Annual and Public Holidays have been taken as listed below and the total number of days are as agreed in the Working Rule Agreement. In some regions the actual day to which the holiday is allocated will vary according to local tradition.

Annual Holidays (21 days)	**Public Holidays (8 days)**	
2 weeks in summer	Good Friday	)
1 week in winter	Easter Monday	)
	Spring	) Summer Period
	May Day	)
	Late Summer	)
	Christmas Day	)
	Boxing Day	) Winter Period
	New Years Day	)

Calculation of hours worked

SUMMER PERIOD
Starting time 8.00am
Finishing time 5.30pm
Lunch period 1.00-1.30pm
Total hours to be worked = 9 hours
   ie overtime is required
30 weeks x 9 hours x 5 days   = 1350 hours
Deduct holidays
10 days summer holiday
        @ 10 x 9   = 90
4 days spring holiday
        @ 4 x 9   = 36
5 days public holiday
        @ 5 x 9   = 45
                   171   171 hours
                              1179 hours

Total hours for Summer Period   1179 hours

WINTER PERIOD
Starting time 8.00am
Finishing time 4.30pm
Lunch period 1.00-1.30pm
Total hours worked = 8 hours
22 weeks x 8 hours x 5 days   = 880 hours
Deduct holidays
7 days winter holiday
        @ 7 x 8   = 56
3 days public holiday
        @ 3 x 8   = 24
                   80   80 hours
                              800 hours

Total hours for Winter Period   800 hours
Total hours worked for 1 year period   1979 hours

ALLOWANCE FOR SICKNESS
Deduct:
  8 days' sickness @ 8½ hrs (average)   68
  (5 days' average + 3 days not payable)

Total hours actually worked in 1 year (Basic hours) =   1911 hours

Inclement weather

The time lost for inclement weather will vary according to the type of work, season of the year, geographical area etc. It is usual to use an average allowance and make any adjustment necessary for exceptional situations in the project overheads.

Say the time lost due to inclement weather is 2%. ie approximately 38 hours

Actual hours worked = Basic	1911 hours
LESS inclement weather	38 hours
	1873 hours

Cost calculation

Guaranteed minimum wages and emoluments (Annual Costs)
(Note: All figures quoted are those *current* as at 27 June 1983.

(a) **Basic wages**

Hours worked per year	1911 hours
LESS time lost for inclement weather: allow 2%	38 hours
Therefore actual hours =	1873 hours

Weekly rate of pay at 27 June 1983 is:
Basic Rate £84.04½ per 39 hour week
Guaranteed Minimum Bonus
Contractors' Bonus Allowances
(or if greater than minimum) £13.45½ per 39 hour week
£97.50 per 39 hour week

Therefore hourly rate of pay is £ 2.50

Cost of all in wage is: 1873 x £2.50 = £4682.50

Additional emoluments

(b) **Inclement weather**
Payment will be made at minimum rates for time due to inclement weather.
Time assumed lost is 38 hours
Therefore cost of inclement weather time is:
38 @ £2.50 = £ 95.00

(c) **Non-productive overtime**
Overtime rates are paid for time worked in excess of the normal working hours.

This allowance is for overtime worked as normal practice; the cost of any additional special overtime will be dealt with separately in the project overheads.

The overtime allowance is calculated as follows:

(i) SUMMER PERIOD

Day	Mon	Tues	Wed	Thurs	Fri	Sat	Sun	Total
Hours worked	9	9	9	9	9	—	—	45
Hours to be paid at flat time rates	8	8	8	8	7	—	—	39
Hours to be paid at time and a half	1	1	1	1	2	—	—	6
Hours to be paid at double time	—	—	—	—	—	—	—	—
Total non-productive time	½	½	½	½	1	—	—	3

30 weeks @ 3 hours per week = 90 hours
   LESS

10 days summer holiday   @ ½ x 8 = 4
                                  @ 2 x 1 = 2

4 days spring holiday   @ 3 x ½ = 1½
                                  @ 1 x 1 = 1

5 days public holiday   @ 4 x ½ = 2
                                  @ 1 x 1 = 1

                                            11½   = 11½ hours

Total for Summer Period   78½ hours

(ii) WINTER PERIOD

Day	Mon	Tues	Wed	Thurs	Fri	Sat	Sun	Total
Hours worked	8	8	8	8	8	—	—	40
Hours to be paid at flat time rates	8	8	8	8	7	—	—	39
Hours to be paid at time and a half	—	—	—	—	1	—	—	1
Hours to be paid at double time	—	—	—	—	—	—	—	—
Total non-productive time	—	—	—	—	½	—	—	½

67

22 weeks @ ½ hour per week	= 11 hours
Less	
7 days winter holiday say one Friday	= ½
3 days public holiday say	= nil
Total for Winter period	= 10½ hours
Total **Summer**	= 78½ hours
Winter	= 10½ hours
Total for whole year	= 89 hours

Therefore, cost of non-productive overtime:-
(Basic Rate per 39 hour week = £84.04½ ÷ 39 = £2.15½)
Therefore overtime allowance = £2.15½ x 89   = £191.80

(d) Sick pay
Assumed number of payable days lost due to sickness is 5.
Statutory rate of payment is £7.00 per day
    Therefore, cost of sick pay is 5 x £7.00   = £35.00

If a private insurance scheme is used instead of a statutory scheme the cost of premiums should be included here instead of the statutory allowances.

(e) Trade supervision
The number of trades foremen to operatives will vary from company to company and in accordance with the needs of a project.

    Assume: (i) that there is one trade foreman for every eight tradesmen;
                    (ii) that half his time is spent working and half on supervisory duties;
                    (iii) that his rate of pay is £0.20 per hour above the trade rate.

Hourly cost for the gang is

1 trades foreman	@ £2.50 + £0.20	=	£2.70
8 tradesmen	@ £2.50	=	£20.00
			£22.70

Allowing for supervision time, the effective hourly cost of the gang is:
        9 ÷ 8.5 x £22.70   =   £24.03½

Therefore, effective hourly cost per operative is:
        £24.03½ ÷ 9   =   £2.67/hr

Therefore, additional hourly cost of supervision is:
        £2.67—£2.50   =   £0.17/hr

Hours worked per year = 1911

Therefore, cost of trade supervision is:

        = 1911 x £0.17   = £324.87

This will vary from firm to firm; it is normal to use a firm's individual arrangements in basic calculations and adjust in the project overheads (if necessary for a particularly project).

(f) **Working Rule Agreement allowances**
Operatives exercising special skills or working in particular circumstances are entitled to special allowances under the Working Rule Agreement. Examples include discomfort, inconvenience or risk, continuous extra skills or responsibility, intermittent responsibility, tool allowances, special provisions or servicing of mechanical plant, storage of tools and clothing. The amount allowed is, therefore, variable according to responsibility and skills. The following is used as an example only.

Say allowance for Working Rule Agreement Allowances received is £2.50 per week. Number of weeks worked during year is 52—5 = 47

    Therefore, cost of WRA allowance is:-
        47 x £2.50   = £117.50

Overheads in employing labour

(g) Training

Many companies have their own training departments and the costs are normally covered in head office overheads or project overheads. However, the Construction Industry Training Board (CITB) has to be supported by the building industry and the operating costs are recovered by way of a levy of rates. The annual levy rates are subject to approval by the Manpower Services Commission and the Secretary of State for Employment. The following levy rates were applicable as at 5 April 1982.

	OCCUPATION	Rate 1982 per annum
1.	Managerial, technical, supervisory	£39
2.	Building crafts	£71*
3.	Roof sheeters, roofing felt fixers	£55
4.	Specialist building operatives	£45
5.	Mechanical engineering services crafts	£85
6.	Scaffolders	£54
7.	Plant mechanics	£25
8.	Miscellaneous craftsmen	£20
9.	Skilled building and civil engineering workers	£25
10.	Labourers	£18*
11.	Demolition workers	£45
12.	Fencers	£30
13.	Steeplejacks	£45
14.	Demountable partition erectors	£45
15.	Electrical engineering services crafts (1981 rate) (— new 1982 rate to be agreed)	£65

Note: Items 2 and 10 marked * are the normal rates applied to craft operatives and labourers.

Labour only services
Where labour-only services are used, the levy is based on a percentage of payments made for labour-only at a rate of *2%*. Firms whose payroll plus payments for labour-only services totalled less than £15,000 during the year ended 5 April 1982, are excluded from payment levy.

Each company will vary in the amount actually paid to the CITB therefore for this example a percentage of 1½% has been used. (see summary).

(h)   National Insurance/Social Security (Contracted in)
Employer's contribution of 11.45% (as at 1 August 1983 for National Insurance including surcharge of 1% (based on average weekly earnings of say £100.00) say   = £11.45

Therefore, cost per year is 52 x £11.45   = £595.40

This is adjusted according to the firm's average weekly earnings per man, all in accordance with Social Security Acts 1975 as amended by National Insurance Surcharge Act 1976. (**Note: the surcharge may be further amended by Parliament**).

(i)   Holidays (including Death Benefit Scheme)
(i) Cost of annual holiday stamp   £10.85
(including contribution of 65p towards death benefit)

Therefore cost per year is 47 x £10.85   = £509.95

(ii) Paid Public Holidays

£84.04½ (Basic Rate) + £13.45½ (GMB) x 8 (days) x 8 hrs ÷ 39 hrs   = £160.00

(j) Severance pay and sundry costs
For the following indeterminate factors allow 2% on labour costs:

- (i)   Severance pay
- (ii)   Loss of production during notice
- (iii) Absenteeism (the cost of National Insurance, pensions and holidays with pay being spread over a smaller number of working hours than the normal conditions assumed).
- (iv) Abortive insurances (paying stamps for operatives who work on Monday in order to have cards stamped but who are absent subsequently).
- (v)   Allow for additional holidays and nationally proclaimed holidays.(see summary)

The percentage allowed will vary from firm to firm according to experience.

(k) Employers Liability and Third Party Insurance
For Employers Liability and Third Party Insurance, allow £2.00% on labour (see summary). This percentage will vary according to the firm, the insurance company, the company's insurance record, and type of work and size of the contract.

Summary   **£**

(a)	Basic rate and guaranteed minimum bonus	4682.50
(b)	Inclement weather	95.00
(c)	Non-productive overtime	191.80
(d)	Sick pay	35.00
(e)	Trade supervision	324.87
(f)	Working rule allowance (extra payments)	117.50
		5446.67
(g)	CITB training contribution @ 1½% on £5446.67	81.70
(h)	National Insurance (11.45% on average earnings of £100)	595.40
(i)	Holidays (i) Annual	509.95
	(ii) Paid Public Holidays	160.00
		6793.72
(j)	Severance payments @ 2% on £6793.72	135.87
		6929.59
(k)	Employers liability @ 2% on £6929.59	138.59
	Annual cost per operative	7068.18

Hourly rate

Number of productive hours worked in year (after deduction of time lost for inclement weather) is 1873 hours.

Therefore, the all-in hourly rate is:-

Annual cost per operative ÷ Actual hours worked

= £7068.18 ÷ 1873   = £3.77/hr

7.04 RECEIPT, ANALYSIS AND SELECTION OF QUOTATIONS

Quotations from suppliers and sub-contractors will be recorded on their receipt in the appropriate quotations register. These registers will also record any discrepances and divergences from the contractor's original enquiry and note any other special features concerning the quotation received.

In selecting the quotation to be used in establishing the net unit rates the estimator will consider:

- the completeness of the quotation;
- the consistency of the rates provided;
- the extent to which additional costs must be considered to overcome discrepancies and divergences. If these are substantial, a decision on the quotation to be used may have to be postponed until *Stage 2,* when such items are costed;

CALCULATING THE 'ALL-IN HOURLY RATE'

REVISED AS AT ....

TABLE A

Text reference		CRAFT OPERATIVES			LABOURERS	
		Omit/Hours	Rate	Amount	Rate	Amount
	Hours per Annum					
	SUMMER PERIOD					
	....weeks x ....days x ....hours =					
	LESS (Holidays)					
	....days Annual Holiday x ....hours =					
	....days Spring Holiday x ....hours =					
	....days Public Holiday x ....hours =					
	WINTER PERIOD					
	....weeks x ....days x ....hours =					
	LESS (Holidays)					
	....days Winter Holiday x ....hours =					
	....days Public Holiday x ....hours =					
	LESS (Sickness and allowances)					
	....days x ....hours =					
	LESS (Inclement weather)					
	....days x ....hours =					
	(or ......%) =					
	Total hours used in calculation					

		£	£	£	£
	GUARANTEED MINIMUM WEEKLY EARNINGS				
a	Basic rate =				
a	Guaranteed Minimum Bonus =				
	(or Contractors bonus allowance) =				
	Weekly rate per [ ] hour week =				
	Hourly rate = ÷ by hours per week				
	Annual rate = Total hours ...... x £ ...... rate per hour =				
	ADDITIONAL EMOLUMENTS				
b	INCLEMENT WEATHER ...... hours x £ ...... rate per hour =				
c	NON-PRODUCTIVE OVERTIME ...... hours x £ ...... basic rate per hour =				
d	SICK PAY ...... days x £ ...... rate per day =				
e	TRADE SUPERVISION ...... hours x £ ...... additional supervision rate per hour =				
f	WORKING RULE AGREEMENT ...... weeks x £ ...... rate per week =				
	SUB-TOTAL A				
	OVERHEADS IN EMPLOYING LABOUR				
g	C.I.T.B. TRAINING CONTRIBUTION @ ...... % of SUB-TOTAL A =				
h	NATIONAL INSURANCE CONTRIBUTION ...... % of average weekly earnings of ...... x 52 weeks =				
i	ANNUAL HOLIDAY CREDIT ...... weeks x £ ...... per week =				
	PUBLIC HOLIDAY CREDIT				
i	CRAFT OPERATIVE £ ...... (Basic Rate) + £ ...... (G.M.B) x ...... days x ...... hours ÷ [ ] hours =				
i	LABOURER £ ...... (Basic Rate) + £ ...... (G.M.B) x ...... days x ...... hours ÷ [ ] hours =				
	SUB-TOTAL B				
j	SEVERANCE PAYMENTS AND SUNDRY COSTS ...... % of SUB-TOTAL B =				
	SUB-TOTAL C				
k	EMPLOYERS LIABILITY ...... % of SUB-TOTAL C =				
	ANNUAL COST PER OPERATIVE				
	Hourly 'ALL-IN RATE' = Annual cost per operative ÷ Actual hours worked per Annum				

- whether deliveries of materials or plant or the provision of services by sub-contractors will be made in accordance with the contractor's programme requirements and method statement;
- that the various factors noted in 6.03 of this Code have been satisfied or noted for cost to be established at a later date.

Due consideration of these factors will determine whether the quotation to be used in establishing the net unit rates can be selected at this stage, or whether selection must be deferred until Stage 2 when other costs are determined, and a full reconciliation of the quotation can take place.

7.05 ESTABLISHMENT OF NET UNIT RATES *(STAGE 2)*

Principles

In calculating unit rates for inclusion in the bills of quantities it is essential that careful consideration is given to every factor which may influence the cost of the work. Some of the common factors are given in the following text.

However, the establishment of realistic production standards is a major consideration outside the scope of this Code, but there can be no substitute for comprehensive company data and feedback from previous work of a similar nature to the project being priced.

Unit rates for measured items in the bills (excluding preliminaries) can consist of any or all of the basic elements:

labour;
plant;
materials;
sub-contractors;
overheads (site and head office);
profit.

It is recommended that each element is analysed and estimated separately so that the total cost of each of the elements can be considered by management.

Production standards and other considerations

The records of cost and outputs achieved on similar work from previous projects should be a major source of the information used in estimating. These data may arise from records of resources used on site or from work study exercises to establish standards. It is important to remember that the cost or output depends upon many variables and attention should always be paid to the conditions which prevailed at the time when the particular recorded cost or output was noted and also to consider the levels of incentives which were used to achieve the particular standard. These conditions must be compared carefully with those expected to be encountered on the project under consideration. Differences between the estimated and actual cost or output on previous projects should be analysed and any obvious conclusions noted. Adjustments must then be made to update the estimating data.

When a particular type of work is being considered for the first time there will be no previous cost or output records for guidance. Specialists should be consulted whenever possible and technical information from outside sources may have to be used; in this respect great caution must be exercised.

Proper allowances must be made for the 'learning curve' on the new type of work and for incentive payments, either by increasing the 'all in labour rate' or by using an appropriately modified production standard. It may be necessary to adjust for any Guaranteed Minimum Bonus or contractor's bonus which has been included in the 'all in labour rate'. This can be adjusted later in the project overheads, if necessary.

Labour element

Labour costs are estimated on the basis of the 'all in hourly rates' previously established. It is recommended that 'gang costs' should for some trades be used in preference to individual hourly rates, eg concreting, plastering and similar bulk work, and then converted into a unit rate for inclusion in the bills. However, it is possible to realistically price all trades on an hourly rate basis providing an allowance for attendant ancillary labour is added to the established 'all in hourly rates'.

Many factors will affect the time allowed for an operation or item and careful consideration should be given to each of these factors, so enabling the time allowed to be as accurate as possible.

The drawings, specification and bills of quantity will indicate:

- the quantity of work to be done;
- what allowance is needed for compaction, overbreak, batters, etc;
- quality of finish and standard of workmanship required;
- whether operations are repetitive;
- whether excessive or detailed setting out will be required;
- degree of accuracy and tolerances required;
- whether the design of the work is intricate or straightforward;
- whether any special skills will be needed;
- whether any special construction sequence is necessary;
- whether the operation is likely to be within the experience of existing staff and operatives, or whether special instruction or training will be needed, or whether there will be a need to engage specially trained personnel;
- position on site in which the work occurs;
- accessibility of work;
- height or depth of work;
- any double handling of materials;
- the weight of specific items;
- restrictions in working, eg. secure areas, safety;
- shift times;
- environment, eg. hot/cold/exposed areas of work.

The tender programme and method statement will indicate:

- time available for activities on the site;
- time of the year when work is to be carried out and the likely seasonal conditions to be encountered;
- whether work will be continuous or intermittent;
- any restrictions which might affect normal working;
- degree of inter-dependence of trades and operations;
- facilities available for use by domestic and nominated sub-contractors as items of general attendance;
- pattern of production and the likelihood of building up to maximum possible rates;
- resources needed, ie relative proportions of supervisory, skilled and unskilled operatives required, and recommended gang sizes;
- extent of mechanisation envisaged and method of unloading, storing, handling and transporting materials.

The visit to the site and locality will have shown:

- physical conditions and any restrictions likely to be encountered;
- site layout, operating storage and unloading facilities;
- likely skill, experience and availability of local labour.

The production standards to be used by the estimator in establishing labour costs will, therefore, take into account the company's existing data and experience, and any circumstances associated with the project information, tender programme, method statement or visits to the site and consultants, which indicate that these standards should be modified in some way. The estimator must record any special factors which lead to alteration of production standards to any considerable extent, in the Estimator's Summary and Report for consideration at the adjudication stage.

Materials element

Various matters have now to be considered with respect to the quotation to be used in establishing the net unit rates. In many circumstances it will not be possible to select the quotation to be used until other factors have been taken into account. Consideration should be given and cost should be established for various additional matters associated with materials. These include:

- any specific divergences or discrepancies from the contractor's enquiry in the quotations received from the supplier;
- any minimum delivery requirements and adjustment of cost due to delivery in small quantities;
- trade discounts which should be noted separately and reported at the adjudication meeting, (Note: discounts may or may not be deducted from the materials' cost at this stage. Some contractors maintain that materials costs should be net of discounts, which are summarised in the adjudication report. Others allow the discount to remain in the materials cost but recognise the element when considering the profit mark-up at the adjudication stage);
- waste allowances;
- unloading, storage and distribution costs.

The degree of mechanisation in unloading must be considered to ensure that material deliveries are compatible with the intended method of handling, ie palletted for handling by forklift truck etc. Special equipment should be considered for unloading, ie skips, slings, chains, etc, although such items are more likely to be costed in project overheads than allocated directly against unit rates. The labour costs of unloading and distributing materials should also be considered and allowance made when establishing the total labour requirements of the project. Such labour can either be taken into account when selecting production standards for labour, or can be priced as a project overhead. Items to be accommodated include:

- storage needs and protection;
- size and weight of materials;
- the cost of any special packaging and crates, if these are charged, or the cost of returning them to the supplier;
- any subsidiary fixing materials or temporary materials needed for storage.

The tender programme and method statement will indicate:

- time and rate of delivery required for materials;
- amount to be stored on site, the location and method of subsequent distribution;
- unloading point.

The visit to the site and locality will have shown:

- physical conditions and any restrictions likely to be encountered;
- site layout, operating, storage and unloading facilities.

The allowance made for waste should, wherever possible, be based on experience gained on previous projects. Data given in text books, periodicals and manufacturers' catalogues, should be examined critically and used with caution. The waste allowance should be carefully applied according to the circumstances of the project and previous experience of the material.

The identification and costing of these various factors and considerations will convert the basic cost contained in the quotation into an adjusted cost which will be inserted into the net unit rates.

Plant element

The analysis of quotations received for plant will be set out in the **'Plant quotations register'** and any additional factors to be priced will have been identified. Allowances must now be made for the additional matters associated with the plant. In considering the total costs of plant, decisions must be made concerning the following items:
- the manner in which time related charges and fixed charges will be accommodated, ie delivery, erection and removal charges could be spread across the duration of hire and added to a weekly rate or alternatively shown separately as a fixed charge separate from time related costs in the project overheads;
- rate of production likely to be achieved by the plant, bearing in mind the specific requirements of the project, the season of the year, and in the case of excavation work, the ground and water conditions. In all cases, the tender programme requirements must be considered;
- the continuity can be expected for any item of plant and the likelihood of building up to a maximum production rate should be evaluated;
- average output should be established, making due allowance for intermittent working, site conditions, seasonal effects and maintenance.

In establishing the costs of plant, further allowances must now be made for additional matters associated with the plant. These include:

- any divergence or discrepancy from the contractor's enquiry in the quotation which is being considered;
- delivery, erection and removal charges if applicable;
- fuel costs, if applicable;
- operator costs, if applicable;
- the effect and cost of maintenance and consequent down time of plant;
- any special provisions needed for unloading and loading plant;
- any temporary access roads, hard-standings or temporary works required for the plant;
- any weight restrictions which may affect the plant or its use;
- whether any special insurances are needed for the plant, ie the contractor is responsible for the plant during delivery and erection;
- any consents required from adjacent owners for the use of plant;
- any contractor's attendant labour requirements, ie banksmen;
- safey measures that are required;
- any supporting equipment needed to operate plant, ie crane slings, chains, skips, cages etc, associated with lifting equipment, hoses, breaker's points, etc, associated with a compressor. (These items may be separately priced in the project overheads);
- allowance for damage and repairs chargeable to the contractor;
- minimum hire charges.

The estimator must decide which items will be accommodated in the 'all in rate' for plant, where plant is to be allocated against unit rates and which items are to be allocated in the project overheads.

MECHANICAL PLANT
CALCULATION OF ALL-IN RATES PER HOUR

Ref.No.

Page _____ of _____

PROJECT													

MACHINE (TYPE AND CAPACITY)

TYPE OF COST	DETAILS	£	£	DETAILS	£	£	DETAILS	£	£	DETAILS	£	£
	SOURCE			SOURCE			SOURCE			SOURCE		
HIRE OF MACHINE												
OPERATOR (if not included under 'HIRE')												
FUEL												
LUBRICANTS (Oil and grease)												
SUNDRY CONSUMABLE STORES												
TOTAL ALL-IN COST PER HOUR												

Allocation of costs

When mechanical plant is used only on specific and limited operations, (such as excavation and soil disposal), then there is little difficulty in allocating the costs of the plant to specific items measured in the bills of quantities, taking into account the various factors noted above.

However, when an item of mechanical plant serves a number of trades or operations, (a crane or hoist or a concrete mixer, which is used for concrete work and also brickwork and drainage work), then the allocation of its cost to measured items can only be made on an arbitrary basis. When the cost of an item of plant is more associated with time on site than to specific items of measured work, (ie pumping operations), then such items cannot reasonably be allocated against measured work.

In such circumstances, the cost of such plant could reasonably be included in the project overheads, rather than spread in an arbitrary manner over measured rates.

Domestic sub-contractor element (including labour only sub-contractors)

The analysis of quotations received for domestic sub-contractors will be set out in the **'Domestic sub-contractors' Register'**. This analysis will have identified any further matters which have to be costed by the contractor. Selection of the sub-contractor to be used in the net unit rates may not be possible before such additional costs have been determined.

Allowances must now be made for any additional matters associated with the domestic sub-contractors works. This can include:

- any specific divergence or discrepancy from the contractor's enquiry included in the quotation;
- any allowance for unloading, storage, protection or materials and equipment and transfer of goods from stores to point of work, if this is to be the main contractor's responsibility. The labour costs associated with unloading and distribution of materials should be considered at this time and allowance made when establishing the total labour requirements of the project. Such labour can either be taken into account by an addition to the sub-contractor's quotation or can be priced in the project overheads;
- any general attendance items to be provided by the main contractor. (See Section 9).

In making any such allowances, the contractor must take into account the requirements of the tender programme and method statement and facilities which have already been allocated for the contractor's own works. Any addition to be made to cover the attendance on domestic sub-contractors may be done in several ways:

- by increasing the relevant unit rates of the sub-contracted work;
- by adding a fixed percentage to the whole of the sub-contractor's quotation;
- by making an addition subsequently in the project overheads.

Any discounts offered on the sub-contracted work should be noted separately and reported at the adjudication meeting. (Note: discounts may or may not be deducted from the sub-contractor's quotation at this stage. Some contractors maintain that sub-contractor's costs should be net of discounts, which are summarised in the adjudication report. Others allow the discounts to remain in the cost of the work to be sub-contracted but recognise the element when considering the profit mark up at adjudication stage).

Great care must be taken in assessing sub-contract quotations to ensure that all items have been adequately covered. If labour only sub-contractors are being considered, the cost allowance must take into account all factors associated with the provision of materials by the contractor, and adequate safeguards must be made to control the use and wastage of such materials.

ADDITIONAL READING
1 NATIONAL FEDERATION OF BUILDING TRADES EMPLOYERS.
Builders overheads and labour rates. 1974. pp37.

8.00 Alternative estimating techniques and daywork

8.01 SPOT ITEMS	– *Labour*
8.02 OPERATIONAL ESTIMATING	– Calculation of standard hourly basic rate for labour
8.03 EXTENSION OF BILLS OF QUANTITY	– Incidental costs, overheads and profits – (the percentage addition)
8.04 PRIME COST OF DAYWORK	– *Materials and goods*
– *Definition*	– *Plant*
– *General*	– *Specimen account format*
	ADDITIONAL READING

8.01 SPOT ITEMS

For estimating purposes spot items may be treated in several ways:

- approximate quantities can be taken off and unit rates used to calculate a lump sum estimate for the item;
- the description within the bills can be analysed into its constituent operations and trades and an estimate of the cost made for each;
- when the description within the bill is analysed into constituent operations and found to have a predominant trade, then a gang or operational assessment can be made on a time, plant and material basis so that the overall cost can be calculated.

In all three methods the cost of labour, plant and materials should be separated in accordance with the general principle described for unit rates.

The work should have been inspected thoroughly at the site visit and where necessary, construction method should be established. Adequate allowances must be made for storage, temporary work, including supports, access, double handling, small deliveries, making good and reinstatement. Transport can be included in the item but will usually be included in project overheads.

If bill descriptions are not clear, or if further information or measurements are required, it may be necessary to revisit the site.

8.02 OPERATIONAL ESTIMATING

In some cases it is preferable not to use normal estimating procedure for calculating the unit rates for particular work items contained in a bill of quantity. The estimator may prefer to consider the inter-relationships of the trades involved in the whole operation.

This is frequently the case with reinforced concrete structures, where the cost of placing concrete will depend on the ratio of formwork, reinforcement and concrete, the chosen method of placing the concrete and the total duration of the operation

CASE STUDY	OPERATIONAL ESTIMATING	REF. No.

BASIC DATA

1. 225mm thick horizontal suspended floor slab with and including a power float finish.
   Total 186.75m^3 in 3 floors –   single pour = $\frac{186.75}{3}$ = 62.25m^3

2. 150mm thick horizontal suspended landing with and including a power float finish.
   Total 1.20m^3 in 3 landings –   single pour = $\frac{1.20}{3}$ = 0.40m^3

3. Staircases.
   Total 3.60m^3 in 6 half-flights –   single pour = $\frac{3.60}{3}$ = 1.20m^3

   TOTAL CONCRETE CONTENT IN A SINGLE POUR   = 63.85m^3 (say 64)

LABOUR ALLOWANCE

TOTALS

Gang size – 4 concretors + 1 banksmen = 5 operatives

TIME ASSESSMENT FOR OPERATIONS	hrs.	OVERTIME CALCULATION	hrs.
(a) Preparation time to set up site	0.50	8 hrs @ ordinary time =	8.00
(b) Concrete delivery in 6m^3 loads		3 hrs @ time and half =	4.50
64m^3 = 11 loads at ½ hr intervals	5.50	3 hrs @ double time =	6.00
(c) General work in tidying up; protection; curing; cleaning plant; preparation for power floating	2.00		18.50
(d) Power float operation	5.50		
(e) Clean up; protection; close and secure site	0.50		
(f) Allowed stoppages – lunch; tea; evening*	1.00	18.50 hrs x 5 operatives =	92.50
*N.B. 1 hr unpaid – lunch; tea, evening breaks			
N.B. Travelling time and allowances priced in Project Overheads	*15.00	ALL-IN LABOUR RATE = £3.24 (example only) TOTAL LABOUR COST = £3.24 x 92.50 hrs	

299.70

PLANT ALLOWANCE

(a) Mobile concrete pump(100mm) . . . . £7.00 hr x 6 hrs . . . . £   42.00
    delivery charges . . . . £10.00 each way . . . . £   20.00
    drivers overtime . . . . £ 4.50 hr x 6 hrs . . . . £   27.00   89.00

(b) 2 no. vibrator poker drive units
    (including delivery charges) . . . . £12.00 day each . . . . £   24.00   24.00

(c) 2 no. power float machines with blades
    (including delivery charges) . . . . £10.00 day each . . . . £   20.00   20.00

(d) Aluminium hand tamp (100x50mm RHS) .£5.00 day each . . . . £   5.00   5.00

(e) Sundry allowance for fuel, temporary access
    and covering equipment . . . . £   20.00   20.00

158.00

MATERIALS

64m^3 concrete including waste allowance @ £22.65m^3 1449.60   1449.60   1449.60

TOTAL COST   £   1907.30

ANALYSIS

Labour   £ 299.70 ÷ 64   = £4.68
Plant   £ 158.00 ÷ 64   = £2.47
Materials £1449.60 ÷ 64   = £22.65

NET UNIT RATE =   **£29.80**

rather than the unit cost of placing individual sections or parts of the concrete. From this operational estimating approach the plant, labour and material rates are calculated based upon the total quantity of work and the total time available to perform the task. This technique is best described by example.

This example (see Page 82) shows a typical calculation used in operational estimating taken from an actual case study.

Case study - project details

A four storey office block situated in a commercial area of a busy town centre. The building has a piled foundation with pile caps, ground beams and a reinforced ground floor slab. The main frame and floors are in insitu concrete with the staircases and landings an integral part of the floor construction.

The site is very restricted with the new building independent but between existing structures. The rear of the building is served by a narrow one-way service road and the front elevation is on a main road (also one-way) with a very busy footpath serving the main shopping town centre. The frontage available and main road are not wide enough to allow a crane to operate from outside the site.

To overcome all these constraints a feasibility study and various method statements were considered in detail. From these studies it became obvious that the method and therefore cost of placing concrete would be determined by the physical constraints of the site and the period of time required to erect formwork and reinforcement.

It was obvious that it would be uneconomical to leave a concreting gang resident on site for the total duration of the operation. Therefore, the selected method of construction was to use pumped concrete and construct the main floor, landings and staircase all in one operation utilising labour brought to the site for this specific operation. The example given examines the calculation of costs for this concreting operation.

Before the actual pour could be commenced the preceding formwork, reinforcement, electrical and mechanical services had to be complete. Therefore a full discussion with all parties concerned ie programmer, contracts management and estimator, was an essential part of the 'operational' considerations. It was also important to take into account the time of year as the weather would seriously affect the power float finishes required, (planned time during the summer period).

8.03   EXTENSIONS OF BILLS OF QUANTITY

When all the net unit rates have been completed, the draft bills should be extended and totalled before the additions are made for project overheads and other allowances. Such extensions are normally made with separate sub-totals being produced for the four basic elements of labour, plant, materials and domestic subcontractors. Further sub-totals are usually produced of any discounts contained within the unit rates in respect of materials, plant, sub-contracted items and PC sums. Alternatively these may be abstracted from summaries of material and subcontractor quotations and the **'Schedule of PC and Provisional sums'**.

The extension of the draft bill rates to produce net unit rates should be carried out in a manner to reduce clerical error and also to allow the establishment of various sub-totals relating to elements of the work or trades as necessary.

Reconciliation will now be possible between the resources envisaged on the tender programme and those contained within the estimate. Such information should be entered into the **'Estimator's summary, analysis and report'** for consideration at the adjudication stage.

8.04 PRIME COST OF DAYWORK

Definition

The prime cost of daywork can be defined in many ways and care should be exercised in reading the definition in the tender documents. The example given is based upon the definition generally accepted for use in contracts issued by the Joint Contracts Tribunal. This is not suitable for jobbing work, work carried out during the defects liability period or work carried out separately from the main contract. It is also not suitable for a wholly prime cost daywork contract.

The composition of the total daywork charge will include the following costs:

- labour;
- materials and goods;
- plant;
- incidental costs, overheads and profit, (this addition will vary between labour, plant and materials)

General

Contractors may decide that project and head office overheads are covered in the contract price and should not be allowed in daywork rates. This is mainly true if the daywork to be carried out during the currency of the contract will not result in extension to the contract but other additional costs to project and head office overheads may still have to be considered.

Overhead costs can also be recovered if work additional to the contract is carried out on a daywork basis.

Decisions concerning allowances for profit and head office overheads must be made by each contractor taking into account his own circumstances, method of working and his assessment of the effects of daywork on a particular project. The contractor should assess each contract on its own merits in producing daywork rates and calculating the percentage addition needed.

Labour

The calculation for the prime cost of labour differs somewhat from that of the all-in-rate, various incidental costs, overheads and profit are deemed to be included in the percentage added to the prime cost rate by the contractor. The percentage added must include all other items that the contractor considers are necessary to recover the true cost incurred but not allowed under the definition of prime cost of daywork.

The calculation consists of:

- guaranteed minimum wages and emoluments;
- additional emoluments in respect of the Working Rule Agreement;
- overhead costs in employing operatives;
- incidental costs, overheads and profit − the percentage addition.

Calculation of standard hourly basic rate for labour

The example given is for a craft operative in NJCBI Grade A areas at 27 June 1983.

Guaranteed minimum wages and emoluments

Standard basic rate 52 weeks x 39 hrs		= 2028	
Less Annual holiday 16 x 8 hrs = 128			
5 x 7 hrs = 35			
	163	163	
		1865 @ £2.15½	= £4019.07½
Guaranteed minimum bonus		1865 @ £0.34½	= £ 643.42½

Additional emoluments

Extra payments under the Working Rule Agreement (WRA) for:

— discomfort, inconvenience or risk;
— continuous extra skill or responsibility;
— intermittent responsibility.

To be adjusted as necessary in accordance with WRA
but for this example only, say £2.50 x 47   £ 117.50

                                                                         £4780.00

Overheads in employing operatives

Employers' National Insurance contribu-
tion @ 11.45% (contracted in) as at 1st
August 1983   £ 547.31
Employers' contribution to:
  Annual holiday credits
  47 weeks @ £10.20 (gross)   £ 479.40
  Death benefits
  47 weeks @ £0.65   £ 30.55
  CITB levy, allow for this example say   £ 71.00

                                                                         £5908.26

Standard hours
Standard hours are calculated as:
52 weeks at 39 hours per week   = 2028

Less
16 days annual holiday at 8 hr/day   = 128
5 days annual holiday at 7 hr/day   = 35
7 days public holiday at 8 hr/day   = 56
1 days public holiday at 7 hr/day   =   7
                                          226   226

                                                           1802 hours

Basic daywork rate

Therefore hourly based rate = Annual cost ÷ Standard hours

£5908.26 ÷ 1802 = £3.28 per hr

It should be noted that remaining costs of employing labour incurred by the contractor in his capacity as an employer, other than those contained in the hourly rate, are to be taken into account under incidental costs, overheads and profit and are contained in the percentage addition.

The example given must be subject to re-examination according to the time when, and where the daywork is executed.

Incidental costs, overheads and profits − the percentage addition

The additional percentage adjustment provided for in most forms of contract incorporates the following:

- head office charges;
- site staff, including site supervision;
- the additional costs of overtime (other than that referred to in the notes below*);
- time lost due to inclement weather;
- additional cost of bonuses and all other incentive payments in excess of the guaranteed minimum bonus;
- apprentices study time;
- subsistence and periodic allowances;
- fares and travelling allowances;
- sick pay or insurance in respect thereof;
- Third Party and Employers' liability insurance;
- liability in respect of redundancy payments to employees;
- employers' National Insurance contributions relating to staff or supervisors;
- tool allowances;
- use, repairs and sharpening of non-mechanical hand tools;
- use of erected contract scaffolding, staging, trestles or the like;
- use of tarpaulins, protective clothing, artificial lighting, safety and welfare facilities, storage and the like that may be available on the site;
- any variation to basic rates required by the contractor in cases where the form of contract provides for the use of a specified schedule of basic plant charges (to the extent that no other provision is made for such variation);
- all other liabilities and obligations whatsoever not specifically referred to above nor chargeable under any other section;
- other overheads and profit.

It should be noted that the illustration given relates to a craft operative. Differing rates will apply to labourers, and to plumbing, mechanical services and electrical trades.

In addition to the above list, regard should be paid to the possible inclusion in the percentage addition of:

- supervision for manual work (only), carried out by principals, foremen, gangers, leading hands and similar categories;

---

*The additional cost of overtime, where specifically ordered by the architect/supervising officer, shall only be chargeable under the terms of the prior written agreement between the parties to the building contract.

- travelling costs not included in the percentage addition;
- lodging allowances not included in the percentage addition;
- Working Rule Agreement payments such as for height, exposed work, cranes, cradles, boats, acid resisting brickwork, dirt, inconvenience, discomfort, intermittent responsibility etc;
- special hourly additions under the Working Rule Agreement for certain trades, sub-trades or operations.

Materials and goods

The prime cost of *materials and goods* obtained from stockists or manufacturers is the invoice cost after deduction of all trade discounts but including cash discounts not exceeding 5% and includes the cost of delivery to site.

The prime cost of *materials and goods* supplied from the contractors stock is based upon the current market prices plus any appropriate handling charges.

Any Value Added Tax which is treated, or is capable of being treated, as input tax (as defined in the Finance Act 1972) by the contractor is excluded.

The contractor's percentage addition may include an appropriate allowance for waste, as well as project and head office overheads and profit.

Plant

The charge for plant is normally applied under the 'Schedule of basic plant charges'. The rates in the schedule are intended to apply solely to daywork carried out under and incidental to a building contract. They are **NOT** intended to apply to:

- jobbing or any other work carried out as a main or separate contract; or
- work carried out after the date of commencement of the Defects Liability Period.

The rates in the schedule are basic and may be subject to an overall adjustment to be quoted by the contractor prior to the placing of the contract.

The rates apply to plant and machinery already on site, whether hired or owned by the contractor.

The rates, unless otherwise stated, include the cost of fuel of every description; lubricating oils, grease, maintenance, sharpening of tools, replacement of spare parts, all consumable stores and for licences and insurances applicable to items of plant. They do not include the costs of drivers and attendants.

The rates should be applied to the time during which the plant is actually engaged in daywork. Whether or not plant is chargeable on daywork depends on the daywork agreement in use. The inclusion of an item of plant in the schedule does not necessarily indicate that the item is in fact chargeable.

Rates for plant not included in the schedule* or which is not on site and is specifically hired for daywork, shall be settled at prices which are reasonably related to the rates in the schedule, having regard to any overall adjustment quoted by the contractor in the conditions of contract.

The contractor will need to allow for various additions to the basic plant rates when compiling the daywork rate for plant. These include:

- general overheads, incidental costs and profit;

---

*Items such as small plant and loose tools, non-mechanical based tools, erected scaffolding, staging, trestles and the like are excluded from the schedule of plant items. They together with consumable stores and protective clothing are normally incorporated into the percentage addition added to the labour rate.

SPECIMEN SUMMARY OF ACCOUNT	£	p
1. Total Labour = £ Add % = £ =		
2. Total Materials and goods = £ Add % = £ =		
3. Total plant, Con- sumable stores etc. = £ Add % = £ =		
4. Add VAT (if applicable) £_____		
TOTAL	£____	

- the effects of inflation on the basic plant charges;
- any special costs arising from insurances etc;
- any differences in the rates that will be paid for plant and the rates contained in the appropriate schedule.

Specimen account format

If the definition of prime cost for dayweek is followed the contractor will submit his daywork account on a voucher or 'daywork sheet', usually designed by the contractor to suit his own company procedure and administration. Whatever form the 'daywork sheet' may take it must include the following elements: labour (including workmen's names), materials and good, plant, consumable stores and services, and be duly signed by the parties to the contract.

A summary of the 'Account' is shown above.

ADDITIONAL READING

1 NATIONAL JOINT COUNCIL FOR THE BUILDING INDUSTRY. National Working Rules for the building industry.

2 RICS and NFBTE. Definition of prime cost of dayworks carried out under a building contract. 1975.

3 RICS. Schedule of basic plant charges. 1981.

4 BAS MANAGEMENT SERVICES. Estimating the prime cost of dayworks under a building contract.

5 ASHWORTH, A. and SKITMORE, M. Accuracy in estimating. CIOB Occasional Paper No. 27. 1983.

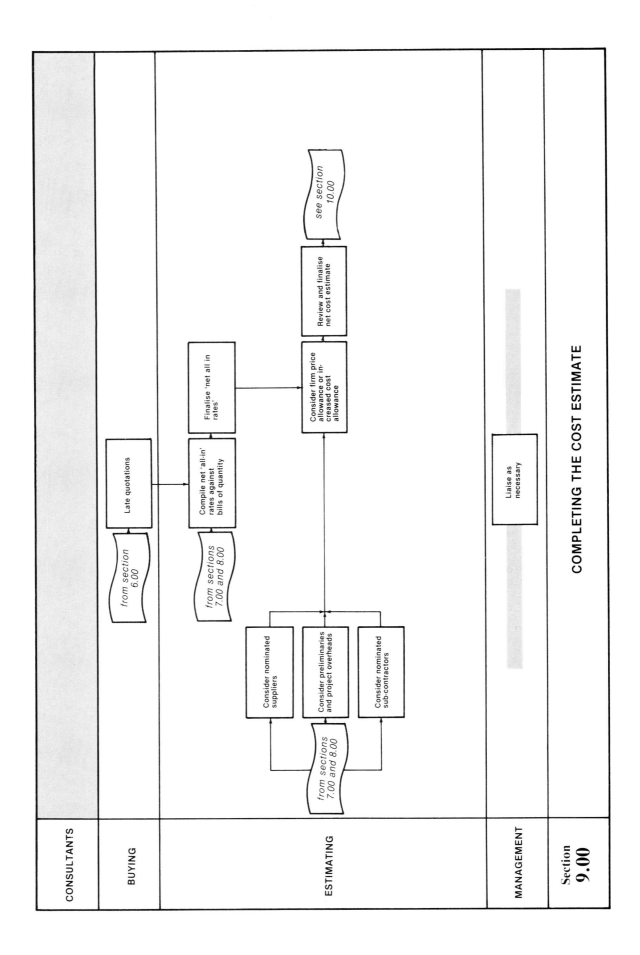

9.00 Completing the cost estimate

9.01	NOMINATED SUPPLIERS	9.06	FLUCTUATING PRICE
9.02	NOMINATED SUB-CONTRACTORS		– *Limited fluctuations* – *Full fluctuations* – *Fluctuations - non-formula method* – *Fluctuations - formula method*

	– *Adequacy of information*
	– *Attendances*
	   – General attendance
	   – Special attendance
	– *Builder's work*
		9.07	ALLOWANCE FOR A FIRM PRICE TENDER

9.03	LATE QUOTATIONS	9.08	CASH FLOW
9.04	PROJECT OVERHEADS *(STAGE 3)* – *Method of estimating* – *Main elements*	9.09	REVIEW AND FINALISE THE COST ESTIMATE
9.05	PRELIMINARIES AND CONDITIONS OF CONTRACT		ADDITIONAL READING

9.01 NOMINATED SUPPLIERS

A list of nominated suppliers will be produced at the Project Appreciation Stage (see Section 5) when PC and Provisional Sums are listed in the **'Schedule of PC and provisional sums'**.

The bills of quantity will normally measure all items associated with the delivery and subsequent inclusion in the works of materials supplied by nominated suppliers. However, it is necessary to ensure that materials are delivered to site in a manner compatible with the contractor's unloading and distribution systems, and method of construction. This may involve additional costs which must be allowed for in the estimate. A nominated supplier for facing bricks may, for example, have allowed for traditional handling, whereas the contractor may wish to off-load by forklift truck. In this case the contractor must allow for the additional cost of delivery using pallets and also for any deposits which are charged on the pallets.

Where the project information is unclear concerning the details associated with a nominated supplier, the estimator should note any particular requirements concerning that supplier in his report for further consideration at the adjudication meeting.

The estimator must also ensure that any fixing items associated with materials provided by a nominated supplier are adequately described and measured in the items to be priced. Any discrepancies concerning fixings, eg plugs, screws, brackets, bolts etc, or ambiguity over the responsibility for supply of these items must be clarified.

Discounts must be listed in respect of goods to be supplied by nominated suppliers and these sums will be noted for consideration at the adjudication stage.

Suppliers may deliver their materials in re-usable crates or other packaging and the contractor may be required to return such items to the supplier. Due allowance must be made for the collection, storage, handling and subsequent dispatch of such items back to the supplier.

9.02 NOMINATED SUB-CONTRACTORS

The work to be carried out by nominated sub-contractors may be a significant proportion of the overall cost of a contract. Contractors are frequently given inadequate supporting information to add to the measured items in the bill of quantities regarding such works. The adequacy of the information provided must be carefully investigated, and further particulars requested by the estimator if details are not complete.

Adequacy of information

At tender stage, the estimator must check that the measured items for works which are covered by a PC sum are adequate and that supporting details are available in accordance with the appropriate Standard Method of Measurement. If not it will be necessary to ask for further particulars before the inter-relationships of trades, the tender programme and method statement can be finalised. All too often this is a neglected area and the estimator must ensure that the tender programme reflects sufficient time for the execution of work which is the subject of a PC sum and that all trades are effectively co-ordinated.

Clarification must be obtained from consultants if there are any doubts as to the adequacy or meaning of the descriptions used.

Attendances

The estimator will also be required to price 'general attendance' and 'special attendance' relating to nominated sub-contractors. Attendance is defined as, *'the labour, plant, materials or other facilities provided by the main contractor for the benefit of the sub-contractor and for which the sub-contractor normally bears no cost'*. The costs associated with attendance are usually built into the main contractor's tender and consequently become a charge against the client. However, the associated risks of attendance are borne by the main contractor.

The main contractor is normally responsible under the main contract provisions for the site establishment and providing attendance and it would be wasteful of space, effort, time and cost if duplicated by sub-contractors.

General attendance

The item for general attendance is intended to be an indication of the facilities which are normally available to sub-contractors where they are provided by the contractor to meet his own requirements.

In assessing any sums to be allowed for general attendance, the estimator must investigate the facilities which will already be provided for the main contractor's use and determine any costs which may arise by the nominated sub-contractor's use of any such facilities. These facilities include:

- use of temporary roads, pavings and paths;
- use of standing scaffolding;
- use of standing power operated hoisting plant;
- use of mess rooms, sanitary accommodation and welfare facilities;
- provision of temporary lighting and water supplies;
- providing space for sub-contractor's own office accommodation and for storage of his plant and materials;
- clearing away rubbish.

Use of temporary roads, pavings and slabs

Allowance must be made for any costs associated with the maintenance of temporary roads, pavings and paths which may be required during the time period allowed by the contractor for his own use. This item will not cover any specific access requirements of a nominated sub-contractor. Such items, for example, a hard standing for a crane should be separately described under 'special attendance'.

Use of standing scaffolding

The contractor should allow for any costs which might arise through the nominated sub-contractor's use of scaffolding which is already erected for the main contractor's use. Any modifications or additional scaffolding required or any extension of programmed time for such scaffolding over and above the time period required by the main contractor should be described and measured as 'special attendance' in the bills of quantity.

Use of standing power-operated hoisting plant

Whilst nominated sub-contractors may use existing hoisting plant if there is spare capacity, any hoisting facilities specifically required should be measured under 'special attendance'.

Use of mess rooms, sanitary accommodation and welfare facilities

Assessment must be made of the accommodation needed for the operations of nominated sub-contractors over and above the requirements of the contractor. Allowance must also be made for any servicing and cleaning of such facilities which are shared with the contractor.

Provision of temporary lighting and water supplies

The estimator must establish requirements for general lighting needed to comply with safety requirements and for the execution of the works during normal working hours. Adequate allowance must also be made for water points needed for the construction of the works. This may mean the simultaneous provision of such services in other areas of the building over and above the requirements of the main contractor.

Special lighting requirements and power needs should be measured and priced under 'special attendance'. Specific water requirements for testing or associated with commissioning of plant should be measured under 'special attendance'.

Providing space for sub-contractor's own office accommodation and for storage of his plant and materials

The estimator should note that only space is required and that cover in the form of a shed is not a requirement. The assessment of total space requirements must be borne in mind when finalising the method statement and site layout.

Clearing away rubbish

The disposal of waste and other rubbish from an agreed collection point involving labour, containers and haulage must be assessed. Abnormal items of rubbish should be measured separately under 'special attendance'.

Special attendance

Other specific attendances which do not fall under the category of 'general attendance' should be specifically measured in the bills of quantities as 'special attendance'. Items to be measured include:

- special scaffolding or scaffolding additional to the contractor's standing scaffolding;

- the provision of temporary access roads and hardstandings in connection with structural steelwork, pre-cast concrete components, piling, heavy items of plant and the like;
- unloading, distributing, hoisting and placing in position, giving in the case of significant items the weight and/or size;
- the provision of covered storage and accommodation, including lighting and power thereto;
- power supplies giving the maximum load;
- any other attendance not included in 'general attendance' or listed above.

Special scaffolding

In order to price this item, the estimator must be given precise details concerning the scaffolding requirements. Such information should define clearly the height in stages of the scaffolding, indicate the extent of boarded platforms and any alteration and adaption that will be required. If such information is not available and descriptions are inadequate, the estimator should seek further instructions from the consultant.

The estimator should also make due allowance under this heading for any adaption or alteration to standing scaffolding or for any extension to the time period, providing such items are described and measured in the bills of quantities.

The provision of temporary access roads and hardstandings

Where any specific requirements are described, these must be taken into account with the contractor's own needs and any additional temporary provision allowed for.

Unloading, distributing, hoisting and placing in position

This item may also include some intermediate storage requirement and due allowance should be made for this. It is essential that particulars are stated of the size or weight of materials to be handled to enable the estimator to reasonably assess costs and identify the appropriate mechanical aids. In the case of heavy units, eg pre-cast cladding, it will also be necessary to be informed of the delivery rate and also if any specific stacking facilities are required for storage on site.

Sufficient information must also be provided to identify any distribution requirements, as opposed to hoisting and stacking. In assessing the cost of such specific facilities, the estimator must bear in mind the extent of use of existing mechanical hoists and must ensure that sufficient hoisting capacity is available to meet all needs.

Provision of covered storage and accommodation including lighting and power

Under 'general attendance' the contractor is required to provide space for nominated sub-contractors to erect their own facilities. Under this item the main contractor will be required to provide, erect and maintain accommodation and provide lighting and power as stipulated. The size of hutting required should be stipulated and the period required stated. Any special requirements, ie racking or other services should also be defined.

Power supplies giving the maximum load

Any special power requirements, including power for testing of systems, should be clearly measured for pricing purposes. Any reference to power supplies should state whether single or three phase supplies are required and the maximum demand level should be taken into account. The estimator should ensure that any descriptions for fuel or power for such testing purposes are clearly specified, giving the quantity necessary to fulfil the tests and also the precise specification of the power needs.

Maintenance of specific temperature or humidity levels

Any specific requirements for controlling temperature or humidity must be clearly measured, stating temperature/humidity required and the time period that the contractor must provide these services. The requirement should also state if the permanent services in the building can be used for this purpose.

Any other attendances

Under this heading, the estimator may be required to price specific attendance or supply materials on various trades. This could include the provision of bedding material for roof tiles, or floor tiles. Other items such as specific cleaning operations, and the removal of masking tape used by sub-contractors should also be defined. Such items should be clearly measured and in the event of any inadequacies or ambiguities, the estimator should refer to the consultants for further instructions.

Builder's work

Items of builder's work in connection with works by nominated sub-contractors will be measured in accordance with the requirements of the Standard Method of Measurement. The estimator will then be required to establish unit rates for the measured items in accordance with the principles previously described. The following points should be borne in mind:

- the extent to which builder's work items are shown on the tender drawings. It may be necessary to ask for further particulars before pricing can be completed;
- the requirement for any specialist work or plant to carry out the builder's work required;
- the rate and timing at which the builder's work should be carried out;
- the continuity to be expected whilst carrying out the builder's work.

9.03 LATE QUOTATIONS

Any late quotations that have been received for the supply of materials, hire of plant or work of sub-contractors, should be entered on the appropriate record sheet and comparison made with previously received quotations. If any of the late quotations offer more satisfactory prices, note should be made of these items in the estimator's report, so that adjustment can be made at adjudication stage.

9.04 PROJECT OVERHEADS *(STAGE 3)*

During the preparation of the cost estimate, various items will have been noted for future inclusion in the project overheads. These must now be brought forward for consideration and noted in the various sections of the **'Projects Overheads Schedule'**. (An example of this Schedule is given on Pages 96 to 107). In addition, other items which have not been included in unit rates must now be considered and costs established.

PROJECT OVERHEADS SCHEDULE

REFERENCE NO.:

PROJECT INFORMATION

Project description:	Client:

Project address:	Tender date:
	Commencement date:
	Completion date:

Dates of co-ordination meetings:	Date of adjudication meetings:

SUPPORTING DOCUMENTS AND INFORMATION — Tick if available

- Preliminary/Tender Enquiry Form
- Site Visit Report
- Schedule of PC sums and provisional sums, daywork and contingencies
- Tender programme
- Method statement

ESTIMATORS NOTES

Value related items may be priced in the summary of the Estimator's Summary, Analysis and Report eg. water charges; insurances; performance bonds

SUMMARY		LABOUR	PLANT	MATERIALS	SUB-CONTRACTOR	TOTAL
A	**TIME RELATED COSTS**					
1	Employers requirements					
2	Management and staff					
3	Site accommodation					
4	Attendant labour and miscellaneous items					
5	Miscellaneous labour costs					
6	Facilities and services					
7	Temporary works					
8	Mechanical plant					
9	Non mechanical plant					
10	Sundries					
11	Contract conditions					
	TOTAL A £					
B	**FIXED COSTS**					
1	Employers requirements					
2	Management and staff					
3	Site accommodation					
4	Attendant labour and miscellaneous items					
5	Miscellaneous labour costs					
6	Facilities and services					
7	Temporary works					
8	Mechanical plant					
9	Non mechanical plant					
10	Sundries					
11	Contract conditions					
	TOTAL B £					
	TOTAL A + B £					

1 EMPLOYERS REQUIREMENTS	No.	£	Wks.	LABOUR	PLANT	MATERIALS	SUB-CONTRACTOR	TOTAL
A TIME RELATED COSTS								
Accommodation								
Architect								
Engineer								
Clerk of Works								
Others								
Telephone								
rental								
calls (if not covered by Provisional Sum)								
Testing and samples								
Other requirements								
(Note: lighting and heating taken in Section 6)								
TIME RELATED COSTS TO SUMMARY		A........£						
B FIXED COSTS								
Erection and dismantling employers site accommodation								
Fitting out and decorating								
Furniture and fittings								
Technical and surveying instruments								
Transportation to and from site								
Telephone installation								
removal								
making good								
Site notice and name board								
Other requirements								
FIXED COSTS TO SUMMARY		B........£						

2 MANAGEMENT AND STAFF	No.	£	Wks.	LABOUR	PLANT	MATERIALS	SUB-CONTRACTOR	TOTAL
(Costs directly charged to the project)								
A TIME RELATED COSTS								
Management — Contracts manager								
Site manager								
Trades supervision — General foreman								
Services engineer								
Foreman								
✶ Trades foreman								
✶ Ganger								
(✶ not in unit rates)								
Engineering — Engineer								
Assistant engineer								
Programming and production control — Programmer								
Assistant programmer								
Production engineer								
Bonus surveyor								
Technical assistant								
Quantity surveying — Senior quantity surveyor								
Quantity surveyor								
Assistant quantity surveyor								
Support staff — Clerk/typist								
Storeman								
Checker								
Timekeeper								
Safety officer								
Watchman/security								
Canteen staff/welfare								
First aider								
Others								
TIME RELATED COSTS TO SUMMARY A........£								
B FIXED COSTS								
FIXED COSTS TO SUMMARY B........£								

3 SITE ACCOMMODATION	DESCRIPTION	£	Wks.	LABOUR	PLANT	MATERIALS	SUB-CONTRACTOR	TOTAL
A TIME RELATED COSTS								
Contractors requirements								
Management								
Trades supervision								
Engineering								
Programming and production control								
Quantity surveying								
Support staff								
Stores								
Canteen/Welfare								
Toilets								
Drying room								
First aid								
Workshops								
Others								
Sub-contractors								
Servicing								
Maintenance of accommodation								
Rates (if not provisional sum)								
(Note: lighting and heating taken in Section 6)								
TIME RELATED COSTS TO SUMMARY		A........£						
B FIXED COSTS								
Erection and dismantling site accommodation								
Fitting out and decorating								
Furniture and fittings								
Transportation to and from site								
Other requirements								
FIXED COSTS TO SUMMARY		B........£						

4	ATTENDANT LABOUR AND MISCELLANEOUS ITEMS	No.	£	Wks.	LABOUR	PLANT	MATERIALS	SUB-CONTRACTOR	TOTAL
A	TIME RELATED COSTS								
	(Items not included in unit rates)								
	Unloading and distribution of materials								
	Attendance upon domestic sub-contractors								
	Drivers								
	Fitters								
	Scaffolders								
	Setting out								
	Progressive cleaning of new buildings								
	TIME RELATED COSTS TO SUMMARY A........£								
B	FIXED COSTS								
	Cleaning building on completion								
	Reinstatement works								
	FIXED COSTS TO SUMMARY B........£								

5	MISCELLANEOUS LABOUR COSTS	No.	£	Wks.	LABOUR	PLANT	MATERIALS	SUB-CONTRACTOR	TOTAL
A	TIME RELATED COSTS								
	Guaranteed time								
	Pensions (where contracted out)								
	Allowances daily fares								
	daily travel								
	weekly subsistence								
	additional bonus								
	Abnormal overtime								
	Attraction money								
	Special severance payments								
	Exceptionally inclement weather								
	Special sick pay provisions								
	Add % for insurances; etc.								
	TIME RELATED COSTS TO SUMMARY A........£								
B	FIXED COSTS								
	FIXED COSTS TO SUMMARY B........£								

6 FACILITIES AND SERVICES	No.	£	Wks.	LABOUR	PLANT	MATERIALS	SUB-CONTRACTOR	TOTAL
A TIME RELATED COSTS								
Telephone — rental								
calls								
Administration — office equipment								
expenses								
stationary, postage								
Safety, health and welfare — cleaning materials								
mess/canteen subsidy								
others								
Cleaning offices and welfare facilities								
Power and lighting — consumption								
surcharges								
generators								
transformers								
leads								
consumables								
Fuels — heating								
plant (not in unit rate)								
gas								
others								
Security — charges								
Rubbish disposal during course of construction								
Drying out of buildings								
Contractors testing and samples								
Seasonal building								
Others								
TIME RELATED COSTS TO SUMMARY	A........£							

6	FACILITIES AND SERVICES	No.	£	Wks.	LABOUR	PLANT	MATERIALS	SUB-CONTRACTOR	TOTAL
B	**FIXED COSTS**								
	Telephone								
	installation								
	removal								
	making good								
	Administration								
	installation								
	removal								
	making good								
	Safety, health and welfare								
	first aid equipment								
	safety equipment								
	protective clothing								
	Power and lighting								
	connection								
	installation								
	distribution system								
	removal								
	making good								
	Water								
	connection								
	installation								
	removal								
	making good								
	standpipes and tanks								
	hosepipes								
	Temporary drainage								
	connection								
	installation								
	removal								
	making good								
	Fuel storage								
	installation								
	storage								
	removal								
	making good								
	testing								
	Security								
	installation								
	removal								
	making good								
	FIXED COSTS TO SUMMARY		B........£						

7 TEMPORARY WORKS

		No.	£	Wks.	LABOUR	PLANT	MATERIALS	SUB-CONTRACTOR	TOTAL
A	**TIME RELATED COSTS**								
Access roads	maintenance								
	cleaning								
Traffic regulations	attendant labour								
	servicing								
Hardstandings	maintenance								
	cleaning								
Pumping and de-watering	attendant labour								
	servicing								
Compounds	maintenance								
	adaption								
Hoardings, fans	maintenance								
	adaption								
Support existing structures and propping									
Progress photographs									
TIME RELATED COSTS TO SUMMARY		A........£							
B	**FIXED COSTS**								
Access roads	provide								
	removal								
	making good								
Traffic regulations	installation								
	adaption								
	removal								
Hardstandings	provide								
	removal								
	making good								
Pumping and de-watering	installation								
	adaption								
	removal								
Compounds	provide								
	removal								
	making good								
Hoardings, fans	provide								
	removal								
	making good								
Support existing structures and propping									
Contractors notice board									
Setting out materials									
Protection									
Other									
FIXED COSTS TO SUMMARY		B........£							

8 MECHANICAL PLANT

	No.	£	Wks.	LABOUR	PLANT	MATERIALS	SUB-CONTRACTOR	TOTAL
A TIME RELATED COSTS								
Cranes * mobile								
* tower								
Hoists goods								
* passenger								
Personnel transport vans								
cars								
* buses								
Site transportation dumpers								
fork lift trucks								
tractor and trailor								
* dump trucks								
* lorries								
Excavating equipment *								
*								
Mixers * concrete								
* mortar								
silos and bins								
Piling *								
Paving surfacing rollers								
compactors								
vibrators								
Other compressor and tools								
powerfloats and blades								
vibrating tamps								
concrete vibrators; pokers								
(* including driver/operator)								
TIME RELATED COSTS TO SUMMARY		A........£						
B FIXED COSTS								
Transport plant to site								
from site								
Costs in receiving; installing and clearing plant								
Temporary works for plant bases/tracks								
power supplies								
Plant workshop and stores								
FIXED COSTS TO SUMMARY		B........£						

PAGE 9

105

9 NON MECHANICAL PLANT	No.	£	Wks.	LABOUR	PLANT	MATERIALS	SUB-CONTRACTOR	TOTAL
A TIME RELATED COSTS								
Small plant and hand tools								
bar benders and cutters								
saw bench and table								
bench saw								
chain blocks								
ladders; steps								
road barriers; lamps								
warning signs; lamps								
✻ props and trench struts								
✻ trench sheeting								
barrows								
Surveying equipment								
levels								
theodolites								
staffs								
Lifting equipment								
slings; chains								
skips; cages								
special equipment								
(✻ *if not in planking and strutting rate*)								
Scaffolding — MEASUREMENT								
Support								
Birdcage								
Access								
Hoist towers								
Mobile towers								
Gantries								
Covers; tarpaulins								
Others								
Alter, adapt								
TIME RELATED COSTS TO SUMMARY		A........£						
B FIXED COSTS								
Small tools and equipment								
hand drills								
percussion drills								
skillsaws, blades								
grinders, blades								
drain testing equipment								
forks, picks, shovels								
Transport to and from site								
mechanical tools								
non-mechanical tools								
scaffolding								
others								
FIXED COSTS TO SUMMARY		B........£						

PAGE 10

106

10 SUNDRIES

	No.	£	Wks.	LABOUR	PLANT	MATERIALS	SUB-CONTRACTOR	TOTAL
A TIME RELATED COSTS								
TIME RELATED COSTS TO SUMMARY A........£								
B FIXED COSTS								
FIXED COSTS TO SUMMARY B........£								

11 CONTRACT CONDITIONS

	No.	£	Wks.	LABOUR	PLANT	MATERIALS	SUB-CONTRACTOR	TOTAL
A TIME RELATED COSTS								
Defects liability period								
Firm price allowance								
Fluctuations price allowance								
Others								
TIME RELATED COSTS TO SUMMARY A........£								
B FIXED COSTS								
Licences and fees								
Excesses and special provisions								
Others								
FIXED COSTS TO SUMMARY B........£								

Method of estimating

The cost of many items will vary in direct proportion to the contract period and the length of time that staff, plant and equipment will be required on the site. It is normal for costs to be built up on a fixed-charge basis and a time-related basis. For example, in setting up a tower crane and then a recurring cost related to the time the equipment is on site. There will then be a subsequent fixed cost for clearing the equipment on completion. It is advantageous to estimate such items in these two distinct forms. Specific information concerning time and method will be obtained from the programme and method statement.

The total estimate of the cost of project overheads will be transferred from this schedule to the final summary document.

Main elements

The items shown in the example form should not be considered as absolute. Items may be added to or omitted from those listed according to the preferences and policies of individual companies. The form provided illustrates typical items which can be required in project overheads. Various techniques exist to price these items and in some organisations the project overheads may be worked back into unit rates. This is a matter for each company to decide.

Supporting documents and information

Essential information needed to establish project overheads has already been compiled in the **'Preliminary/Tender Enquiry Form'**, the **'Site Visit Report'**, the **'Schedule of PC Sums, Provisional Sums and Daywork'**, the **'Tender Programme and Method Statement'**. In addition, it may be necessary to refer to the tender documents for further information on specific matters. The estimator must ensure that all this relevant information is brought together before the resources needed in project overheads are defined and costed.

The schedule of project overheads

Various groupings of items relating to project overheads are possible according to company preference. The sample forms consider establishing costs on a multi column basis in order to allow the build up of costs in a similar way to unit rates in terms of labour, plant, materials, and sub-contractors. Not all columns are relevant to all sections but this consistent approach and layout allows costs to be established under these major headings. The following are the main groups of items to be considered:

1) *Employers requirements*

    Including all accommodation and other specific requirements that may be required by consultants and clerk of works.

2) *Management and staff*

    This will include all personnel whose costs have not been included in unit rates or head office overheads, whatever their method of remuneration. The costs allowed will normally include all costs associated with employing the staff. Company cars and their running costs can be allowed for under 'Plant' if it is the company's policy to view such items as part of the employee's cost.

    If visiting head office staff are charged for projects as part of the company's policy, the estimator must liaise with management over the extent of such head office services to ensure that due allowance is made in the cost estimate.

3) *Site accommodation*
The hire or purchase of accommodation should be considered under this section. Reference to the method statement and programme will be necessary to determine the intended layout of the site and time period for the various elements of accommodation. Transportation, erection, fitting out and decoration, subsequent dismantling and reinstatement is noted in the fixed cost section of the form. Repetitive items such as maintenance are noted under time related costs.

Cleaning costs and electricity and fuel are noted under item 6) 'Facilities and services'.

4) *Attendant labour and miscellaneous items*
It is sometimes convenient to group together various miscellaneous labour matters under project overheads rather than pricing in unit rates. This is particularly the case when a number of trades require attendance from the main contractor or when dealing with unloading and distribution of materials. A general allowance priced on a time-related basis is often more appropriate than attempting to allocate resources back into unit rates.

Other items such as cleaning the building during the works and final cleaning on completion are also normally priced in project overheads. Attendant labour on sub-contractors can also be priced in project overheads if such items are not included in unit rates. The drivers of general plant can also be priced here if not included in the price of the plant item.

5) *Miscellaneous labour costs*
When unit rates for labour were being examined in Section 7 it was noted that certain costs attributable to the employment of labour were often contained in the project overheads. Some costs relating to labour are not clear at the early stages of preparation of the estimate when the volume of labour required on the project has not been accurately established. These items, which have not been included in the unit rate calculation must now be brought forward for pricing in the project overheads.

The allowance made for travel and fares allowance, subsistence allowance, attraction money and exceptional inclement weather will be based upon information gained from the site visit when an appreciation of the labour availability will have been made. This item, together with attraction money and additional bonus allowances, are critical areas for contractors and the estimator will take considerable care in making his recommendations to management in such matters.

The allowance for special severance payments over and above the requirements of the statutory redundancy requirements, refer to situations where it may be necessary to pay a special terminal bonus to operatives because of particular circumstances associated with the project or its location.

6) *Facilities and services*
The installation, removal of services and any subsequent reinstatement that may be required, together with time related costs and consumable items is dealt with in this section. Other services and facilities include cleaning, office facilities security, rubbish disposal, testing and seasonal building.

7) *Temporary works*
The costs of temporary access, hoardings and such like items are determined in this section. Recurring costs to maintain access etc should be calculated and the costs of final removal and any reinstatement of the area should also be established. The method statement and tender programme will be important documents when establishing the costs of these aspects of project overheads.

8) *Mechanical plant*
The estimator must produce 'all in rates' for many of these items of plant. Drivers or tradesmen, where needed, can either be incorporated as part of the 'all in rate' or alternatively shown under Section 4) 'Attendant labour and miscellaneous items'. Attendant labour, ie operating pumps, hoisting equipment etc together with down time for servicing, should be established by the estimator when considering 'all in rates'.

If quotations have been obtained for the plant in question the estimator must ensure that the procedures outlined in 6.03 are followed and that the quotation used is satisfactory and that extra allowance or attendances are allowed for.

The method statement and programme will give valuable guidance at this stage and indicate the type of plant required and the time period the plant will be required on site. Specific temporary works associated with plant must also be established with allowances made if not already covered under Section 7) 'Temporary works'.

A special electricity supply may be needed for large plant such as a tower crane and as well as the installation and running costs of such equipment the estimator must bear in mind the possibility of surcharges on this cost from the Electricity Board.

9) *Non mechanical plant*
'All in rates' will be needed for some of the plant in this section whilst others may well be regarded as a consumable item and charged fully against the contract. The estimator should consider the question of likely credits against such equipment when establishing the allowances to be included in the estimate.

10) *Sundries*
Any unusaul features associated with the project which do not fall naturally into the defined categories can be priced in this section.

11) *Contract conditions*
Allowances for firm price or fluctuating price are considered in 9.06 and 9.07. Other matters concerning the contract conditions will also be considered under this heading. Some costs are 'value related' and a decision concerning the cost of these items may be delayed until the final summary is considered and final costs are established.

9.05 PRELIMINARIES AND CONDITIONS OF CONTRACT

The Preliminaries section of the bills of quantity and actual conditions of contract should be scrutinised to ensure that all items attracting a monetary value have been included in the cost estimate.

The cost of bonds, insurances and water for the works (where applicable) should not be added until after adjudication if these items are calculated as a percentage of the work value.

9.06 FLUCTUATING PRICE

The method of calculation of fluctuations in costs is usually set out in the appendix to the conditions of contract. Fluctuations in cost can be 'limited' or 'full'. It should be borne in mind that fluctuations in cost can give rise to both increases and decreases in cost and careful consideration is needed to reconcile such changes to the contractor's cost estimate. In addition to the 'full' and 'limited' fluctuations the estimator must consider whether over or under recovery of actual increased costs will occur as a result of the proposals in the conditions of contract for fluctuations.

Limited fluctuations (Statutory items)

This is normally limited to specific changes to rates of contribution, levies and taxes in the employment of labour and in the rates and duties and taxes on the procurement of materials. The extent of fluctuations of such costs are limited and are effectively those which arise under and by virtue of an Act of Parliament.

Full fluctuations

Full fluctuations in cost are intended to provide an equitable means of reflecting changes in cost throughout the duration of a contract. However the estimator must ensure that any shortfall or non recovery items are identified and also any over recovery highlighted. Due allowance should be made in the project overheads for any adjustment that is needed. Adjustments are limited to national increases and do not, therefore, reflect market forces.

Fluctuations – non formula method

The contractor can find no adjustment is made on:

- the cost of calculating fluctuations;
- overheads and profit;
- consumable stores and small tools;
- project overhead costs;
- standing bonuses;
- bonus schemes not complying with the Working Rule Agreement;
- materials and plant not appearing on the list of materials and basic prices;
- in certain cases, fuels.

Inadequate adjustment can occur on:

- others employed on the site;
- employer's pension scheme contributions where contracted out;
- materials purchased in small loads;
- materials not forming a significant part of the work.

These lists are not all-embracing and will vary according to conditions of contract. Specific examination should be made on each contract to highlight all areas where adjustment will not occur or will be restricted. This is particularly important when under recovery will occur.

Labour

The estimator must establish the extent to which the contractor will recover increased costs in relation to:

- increased rates of wages promulgated by a recognised wage fixing body;
- other emoluments such as bonus and overtime and such expenses as may be determined in any Working Rule Agreement;
- increases in holiday stamps;
- increase in the cost of employer's liability insurance, third party insurance and any contribution, levy or tax paid by the contractor as an employer.

Full recovery of increased cost may not be made on all increased wages, bonuses, expenses, contributions, levies and taxes and any shortfall must be calculated and due allowance made in the **'Schedule of Project Overheads'**.

Careful attention must also be paid to such areas as fares and transport costs for operatives and also to pension provisions and any under-recovery should once again be allowed in the **'Schedule of Project Overheads'**.

Market pressures can frequently bring about increases in costs of labour which are not recoverable.

Materials

Fluctuations in cost are normally made by reference to a list of materials and basic prices. The responsibility for production of this list varies according to conditions of contract and such lists may be produced by the consultants and form part of the tender documents. Alternatively, the contractor may be required to produce such a list during the preparation of his estimate.

When the list is produced by the consultants the estimator must thoroughly examine the list for omissions. If items are not listed there can be no adjustment for fluctuating costs. If consultants are not prepared to amend the list to add items requested by the estimator then due allowance for fluctuation in cost of unlisted items must be calculated by the estimator and noted in the **'Schedule of Project Overheads'**.

When the list is to be prepared by the contractor the estimator must ensure that it is as full and comprehensive as possible. Items not included in the list should be noted and allowance for fluctuation in cost included in the **'Schedule of Project Overheads'**.

Sub-contractors

The materials and goods of domestic sub-contractors must be included in the list if fluctuations are required. The estimator must ensure that such information is obtained on the quotations received from sub-contractors and that the prices quoted relate to the date of tender under the main contract. The failure to include relevant particulars relating to sub-contractors could penalise the contractor in that with costs increasing he may be unable to receive these increased costs on items relating to sub-contractors' work from the employer. He may still have a liability to reimburse the sub-contractor for such costs. Market pressures can frequently bring about increases in costs which are not recoverable.

The importance of systematic production of the list of materials and basic prices cannot be overstressed.

Fluctuations − formula method

When a formula method is used for adjustment of costs the amount the contractor recovers from the employer is calculated by reference to pre-determined rules. It can be considered as an adjustment in cost which has been deemed to occur by the contractor. Most contractors who operate the formula fluctuations agree that the amounts recovered do give a reasonable recovery of the increased costs that occur.

The estimator should ascertain that the deduction normally applicable when formulae are used for fluctuations calculations, does not result in any short-fall in cost recovery to the contractor. Any under or over recovery expected should be noted in the project overheads for consideration at adjudication stage.

The cost indices upon which the amounts of fluctuations are calculated are based on data collected from British manufacturers. The indices therefore do not reflect the cost changes relating to goods which are manufactured abroad. Transport costs and currency changes are also at risk.

In such situations the estimator must establish the position of fluctuations in costs under the conditions of contract and make assessment of any short-falls or over recovery which should be added to the **'Schedule of Project Overheads'**. The estimator may be required to submit a list of goods which are to be imported with the tender and such a list should form part of the documentation considered at adjudication stage.

9.07 ALLOWANCE FOR A FIRM PRICE TENDER

When a firm price tender is to be given, an assessment must be made of the likely variations in cost during the proposed contract period, including the period for 'acceptance of tender' stated in the contract documents.

Each company must keep records of wage rates and plus rates due under the Working Rule Agreement and bonus earnings. Note must be taken of industrial negotiations and statutory measures which may affect levels of wages, the financial obligations to the employer or any costs that must be taken into account as a result of such negotiations. If details are not agreed at the tender date then it is recommended that discussions with the NFBTE or other appropriate parties should take place requesting guidance as to the likely outcome of such negotiations. Records must also be kept of the cost of materials, plant hire etc. From these records and with detailed discussions with suppliers and manufacturers, the likely trends in prices can be more accurately assessed, rather than making an arbitrary guess at price increases which appears to be the case in many assessments. Consideration must also be given to indirect charges such as tariffs, fuel and freight.

Although it is a very difficult assessment to make, regard must be paid to economic and political situations at home and abroad and the state of both home and world markets. The quotations from the contractor's own sub-contractors must be examined very carefully to ensure that each offer is for a firm price contract and for the period required by the contractor's invitation details and that there are no risks of claims for increased costs.

The assessment of additions for increased costs, together with the decision to submit a firm price tender, will frequently be two of the major decisions in any tender.

9.08 CASH FLOW

Cash is a major resource in construction and its use must be anticipated and managed. Production on construction projects must be planned and the effect upon resources and the return on assets employed determined. Clearly, the return on capital will be influenced by the speed at which it is earned. A simple cash flow diagram must be regarded as a useful aid to decision making and the estimator should ensure that such a diagram is produced for consideration at the adjudication stage.

When a contract has been secured then a fully detailed forecast of cash flow, budget and cost control will be essential but it is not intended to include post contract action within this Code.

The estimator's build-ups will have established the costs of the work and project overheads under the headings of, labour, plant, materials, domestic sub-contractors, PC sums and provisional sums and daywork. In order to produce a cash flow diagram this financial appraisal must be linked to the tender programme. Cash resources must be established for each item on the programme. Once this link between estimate and programme has been established the cash flow position can be calculated. The following headings should be taken into account when assessing the financial commitment to the project and the cash flow forecast:

The contractor's own trading conditions

- the anticipated payment period for own labour, plant, materials, own sub-contractors and suppliers.

Contract conditions

- time period before issue of the first certificate;
- interval between certificates;
- anticipated time of practical completion;
- anticipated time for ending of defects liability period;
- delays between issuing of certificates and receipt of cash;
- level of retentions;
- maximum retention;
- retention during the maintenance period;
- period for final measurement.

Banking conditions

- earning rate of interest;
- paying rate of interest.

Other factors which may also need to be considered include:

- timing of special payments, ie special insurances;
- bonds, including date of release;
- the allocation of allowances for firm price;
- the recovery of fluctuating costs;
- any special weighting that may be incurred on the contract;
- provisions for liquidated damages;
- any purchases of plant or equipment;
- the present financial commitment of the company and effect of the current tender on that commitment;
- company's current turnover and commitment of resources;
- statutory requirements, eg VAT;
- level of project overheads and means of recovery of cost;
- amount of PC and provisional sums;
- company's prior experience of the client and his consultants;
- the addition to be made to the net cost estimate for head office overheads and profit and the manner in which this addition is to be made.

Cash flow calculation will indicate the financial support needed by the contract and the financial contribution to be made to the company. This important matter must be considered at adjudication stage.

9.09 REVIEW AND FINALISE THE COST ESTIMATE

Before the cost estimate is presented to management for adjudication it must be carefully reviewed. The calculations for 'all in rates' must be examined and checks made to ensure that relevant figures and assumptions were used in the build-up of the unit rates.

Each page of the bills must be inspected to ensure that no item or part of an item has been omitted.

The total unit rates must appear reasonable and be correct for the unit of measurement. All calculations should be subjected to an arithmetical check. Particular attention must be given to the major items in each trade and to those items in which the materials specified are unusually costly. Any inaccuracy in the rate build-up will have been magnified and have very serious consequences.

It is equally important to check the sub-contractor related items to see that items of work, plant and material not included in their quotation have been correctly assessed and co-ordinated into the rest of the item and the unit rates reflect these factors.

It is important to ensure that when the time to be allowed for the project is given by the client, it is realistic in relation to the estimated labour cost and number of men assessed as required to be physically working on site.

The construction programme must be reviewed critically to ensure that no major problem has been overlooked and that the method statement represents the best method of working on the project; also that the cost estimate accurately represents in financial terms the true cost of work to be carried out.

The original estimate and final costs of previous projects of a similar nature must be examined and an opinion formed of the likely profitability of the project under consideration.

ADDITIONAL READING

1 JACKSON, D. Estimating aspects of fixed price contracts in *The practice of estimating*. 1981. CIOB.

2 JONES, G. P. An appraisal of formulae used for fluctuations in construction costs. CIOB Occasional Paper No. 9.

3 NEALE, R. H. and EVISON, P. J. A comparison of series 1 and series 2 price adjustment formulae for building contracts in *The practice of estimating*. 1981. CIOB.

4 SHARP, J. A. A. The cost estimate – a need for reconciliation. CIOB Estimating Information Service Paper No. 40.

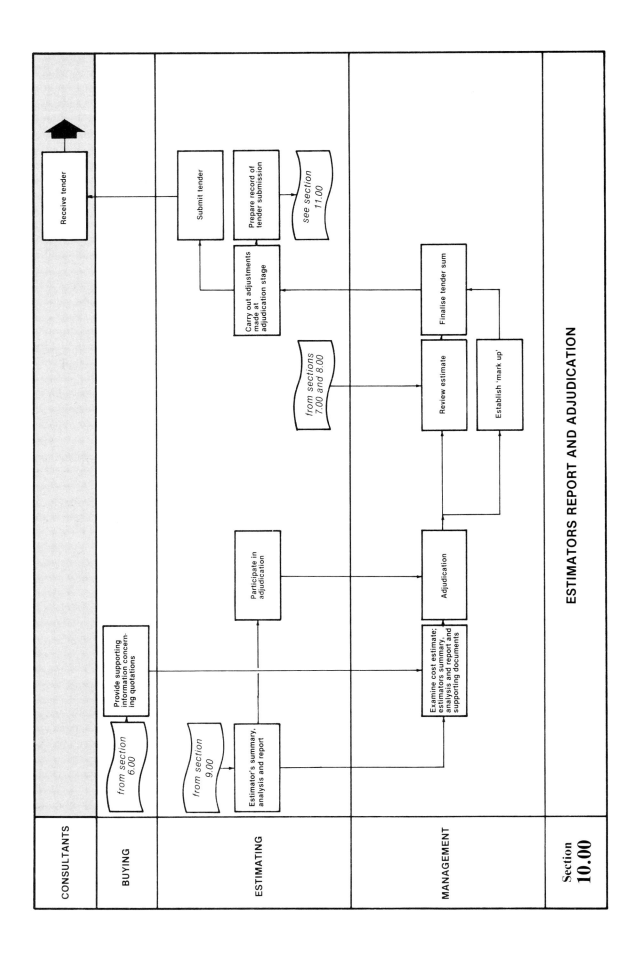

10.00 Estimator's report and adjudication

10.01 ESTIMATOR'S SUMMARY, ANALYSIS AND REPORT	10.02 ADJUDICATION
– *The report and analysis* – *Supporting documents* – *The report form*	– *Tender adjudication meeting* – *Matters to be considered* – *Submission of priced bills of quantity* – *Records*

10.03 SUBMISSION OF THE TENDER
ADDITIONAL READING

10.01 ESTIMATOR'S SUMMARY, ANALYSIS AND REPORT

The estimator's summary analysis and report will bring together all pertinent facts which have influenced the preparation of the estimate for adjudication by management. The objective of this report is to highlight to management the various matters which have been identified as cost significant, where alterations have been made to normal production standards and any special or unusual contract conditions or risks. These matters are described more fully below. An example of a form which can be used for the estimator's report is shown on Page 118. It should be noted that this is only one representation of data which can be considered at this stage. The way in which information is presented will vary according to company policy and preference and the guidelines suggested should not be regarded as mandatory.

The report and analysis

The estimator's report to management should include:

- a brief description of the project;
- a description of the method of construction;
- a note of any actual risks which are inherent in the project;
- the Conditions of Contract;
- any unresolved technical or contractual problems;
- an assessment of the state of the design and the possible financial consequences thereof;
- a note of any major assumptions made in the preparation of the cost estimate;
- an assessment of the profitability of the project;
- any pertinent information concerning market and industrial conditions;
- any need for qualification of the tender or for an explanatory letter;
- the terms of quotations from own sub-contractors which have been included in the estimate;
- the time for which the tender is to remain open for acceptance;
- any information regarding the client, architect, quantity surveyor, consultants or other member of the professional team that should be brought to the attention of management;
- details of other tenderers where known;
- employer's special conditions, ie bond/special insurances/etc.

The report will include various documents which have been produced during the preparation of the estimate and various summaries relating to the cost estimate, including:

ESTIMATORS SUMMARY, ANALYSIS AND REPORT

Ref. No.
Page   of

PROJECT INFORMATION

Project description	Client
Project address	

SUPPORTING DOCUMENTS AND INFORMATION	(Tick if available)
Preliminary/Tender Enquiry Form	
Site Visit Report	
Schedule of PC and provisional sums, daywork and contingencies	
Tender programme	
Method statement	
Schedule of Project Overheads	
Cash flow calculation	

ESTIMATORS COMMENTS

		Page of

A CONTRACT PARTICULARS

FORM OF CONTRACT	

SPECIAL CLAUSES OR CHANGES TO STANDARD FORMS

No.	Notified at time of tender	Further particulars received during tender period

B OTHER INFORMATION

Item	Tender documents	Further particulars received during tender period
Tender date		
Defects liability period		
Date for possession		
Date for completion		
Liquidated and ascertained damages		
Period of delay		
Period of interim certificates		
Period of honouring certificates		
Percentage of certified value retained		
Limit of retention fund		
Period of final measurement and valuation		
Period for issue of final certificate		

C DRAWINGS AVAILABLE

Tender documents	Tick if contract doc.	Further particulars received during tender period

D

ANALYSIS OF BILLS

Bill No.	Trade	Own work Labour £	Plant £	Materials £	Sub-contract £
	Totals to Final Summary				

OF QUANTITY

Page of

ominated contractors	Nominated suppliers	Discounts own work	Discounts nominated items	Provisional sums	Contingencies	Trade totals
£	£	£	£	£	£	£

E		SUMMARY OF PROJECT OVERHEADS		Page of	
	Description	Amount £	Details of adjustment	Add £	Deduct £
A	**TIME RELATED COSTS**				
1	Employers requirements				
2	Management and staff				
3	Site accommodation				
4	Attendant labour and miscellaneous items				
5	Miscellaneous labour costs				
6	Facilities and services				
7	Temporary works				
8	Mechanical plant				
9	Non-mechanical plant				
10	Sundries				
11	Contract conditions				
	TOTAL A				
B	**FIXED COSTS**				
1	Employers requirements				
2	Management and staff				
3	Site accommodation				
4	Attendant labour and miscellaneous items				
5	Miscellaneous labour costs				
6	Facilities and services				
7	Temporary works				
8	Mechanical plant				
9	Non-mechanical plant				
10	Sundries				
11	Contract conditions				
	TOTAL B				
	TOTAL A + B		TOTALS		
	Net adjustment (ADD/DEDUCT)				
			Adjustment		
	Total to FINAL SUMMARY		Net adjustment		

| F | | LATE QUOTATIONS AND ADJUSTMENTS | | | | | | | Page | of |

MAIN CONTRACTOR'S WORK

Bill No.	Quantity	Adjustment	Add			Deduct		
			Labour	Plant	Materials	Labour	Plant	Materials
			£	£	£	£	£	£
Totals to Final Summary								

DOMESTIC SUB-CONTRACTORS' WORK

Bill No.	Quotation used	Amount	Alternative quotation	Amount	Add	Deduct
		£		£	£	£
Totals to Final Summary						

	Page of

G   ESTIMATORS REPORT AND RECOMMENDATIONS

(N.B. Note major cost significant items, risk areas, alterations to normal productions standards)

H   ADJUDICATION NOTES

FINAL SUMMARY

Page of

BILL SUMMARIES (F) (D)	Gross cost	Discount	Net cost	Adjustments	FINAL TOTAL
Labour					
Plant					
Materials					
Domestic sub-contractors					
PC AND PROVISIONAL SUMS DAYWORK AND CONTINGENCY (D)					
Nominated suppliers					
Nominated sub-contractors					
Provisional sums					
Contingencies					
PROJECT OVERHEADS (E)					
Time related costs					
Fixed costs					
DISCOUNTS ADJUSTMENT					

	NET TOTAL	

VALUE RELATED ITEMS (if *not* allowed elsewhere)		
Project insurance		
Third party insurance		
Special clauses/liabilities		
Water for works		
Others		
	FINAL NET TOTAL	
MARGIN { GENERAL OVERHEADS		
PROFIT		
BONDS		
TENDER FIGURE		
Signatures		

- summary of the cost estimate;
- summary and supporting quotations for:
    domestic sub-contractors;
    materials;
    plant;
- tender documents and contract drawings;
- summary of project overheads.

Supporting documents

These items will in turn be supplemented by supporting documents which will provide additional detail to material contained in the various summaries and report including:

- Preliminary Tender Enquiry Form;
- Site Visit Report;
- Schedule of PC sums, Provisional Sums and Daywork;
- Tender Programme;
- Method Statement;
- Schedule of Project Overheads;
- Cash flow calculations.

This information will then be considered by management at the adjudication stage.

10.02 ADJUDICATION

The adjudication of an estimate and its conversion to a tender is the responsibility of management and is a separate commercial function based upon the cost estimate and its supporting reports and documents.

The accountability of the estimator should be limited to the proper preparation of the predictable cost of a project. It must not be considered that the estimator's responsibility is to secure work for his company; that is the responsibility of management.

The adjudication of an estimate into a tender in the construction industry is similar in thought and method to any other manufacturing industry; the need to control a business in financial and production terms is paramount to the success of any business.

Tender adjudication meeting

Although the final decision is for management it is important that those concerned with the planning, management, estimating and buying should be encouraged to communicate the knowledge they have acquired throughout the estimating period as an aid to the decision making.

This may well entail the setting up of two-tiered meetings in larger organisations.

(i) management and directorate: estimate review;
(ii) directorate: commercial decisions and 'mark-up'.

The need for a formal approach should be regarded as fundamental to competent tendering.

An agenda for adjudication meetings is indispensable until such time as orderly logical and methodical thinking becomes second nature.

Matters to be considered

- all substantive items in the estimator's project report;
- contractual commitment;
- parties involved, eg client, architect, quantity surveyor, consultants;
- method statement, planning report and tender programme;
- project overheads and supporting reports;
- manpower requirements, eg managerial, technical, sub-contractors, present workload;
- contract conditions: amendments to standard forms; deletions from standard form appendices:
    - (a) time stipulated or assessed;
    - (b) damages;
    - (c) fluctuations or fixed price;
    - (d) insurances;
    - (e) bond – percentages.
- financial
    - (a) payments;
    - (b) fixed price or fluctuations;
    - (c) non-recoverable expenditure;
    - (d) cash flow;
    - (e) other considerations.
- summary: analysis of net cost estimate;
- qualifications: discuss basis for and wording of any qualifications;
- competitors: knowledge of and past results;
- mark-up: consider:
    - (a) quality of project information;
    - (b) risk;
    - (c) head office overheads;
    - (d) profit;
    - (e) discounts;
    - (f) financial considerations and cash flow;
    - (g) VAT (where applicable) etc.

(Note: VAT may also be considered separately, depending on the type of work and contract requirements).

Submission of priced bills of quantity

At adjudication stage management may alter some of the basic conceptions concerning the project. A change to the basic labour rate will affect all bill items containing labour and, therefore, affect all net rates that have been compiled by the estimator. With computer aided-estimating systems such changes can be easily accommodated at the end of the tender period but often there is insufficient time to rewrite the net rates when manual estimating systems are used.

If priced bills of quantity are to be submitted at the time of tender (or if requested later), management must decide how the difference between the total of the net rates in the draft bills and the agreed tender figure is to be shown. The method chosen will depend on individual preference.

Examples of possible methods are:
- all amendments being made to the respective elements of the tender. (This may mean substantial repricing of the bills of quantity);
- unit rates are increased by an agreed percentage so that the whole difference is included in the measured items;
- the unit rates are left net and the whole difference is included as a percentage on the final summary page;
- the unit rates remain net and the difference is included in the Preliminaries Bill;
- any combination of these methods.

Stage 6 of the *Code of Procedure for Single Stage Selective Tendering* recommends alternative methods of dealing with errors in bills of quantity and these should be noted when submitting tender documents.

Records

It is recommended that a record is kept of all tenders submitted. Particular care should be taken of the adjudication report and all notes and details of the decisions taken at the adjudication meeting should be fully recorded for future use.

10.03 SUBMISSION OF THE TENDER

The estimator must ensure that the procedure set out in the tender documents for the submission of the tender is followed meticulously. The form of tender must be completed as required and signed by an authorised person from the contractor's organisation. Any other declarations such as a fair wages resolution, bond details or no collusion clause must also be appropriately signed.

If alternative proposals have been invited then these should be clearly stated. Any comments regarding the contractor's assessment of the contract period, if this is to be determined by him, should be set out in unambiguous terms.

The *Code of Procedure for Single Stage Selective Tendering* recommends that tenders should be opened as soon as possible after receipt. Submission of priced bills of quantity may be required with the tender but in any event CPSSST recommends that the lowest tenderer should be invited to submit priced bills of quantity within four working days of the opening of the tenders. The estimator must ensure that bills are properly written up, extended and totalled so that submission can be made at the appropriate time if required.

ADDITIONAL READING

1 NATIONAL JOINT CONSULTATIVE COMMITTEE. Code of procedure for single stage selective tendering. 1977. RIBA.

2 TASSIE, C. Aspects of tendering: converting a net estimate into a tender in *The practice of estimating*. 1981. CIOB.

3 DAVIES, F. A. W. Preparation and settlement of lump sum tenders for building works. Ibid.

11.00 Action after submission of a tender

11.01 ASSESSING TENDERS AND NOTIFYING RESULTS	– *Contract documents* – *Distribution of information* – *Cost information* – *Visit to the site*
11.02 ADJUSTMENT OF ERRORS	– *Final reconciliation*
11.03 ACTION WITH A SUCCESSFUL TENDER	11.04 ACTION WITH AN UNSUCCESSFUL TENDER
	ADDITIONAL READING

11.01 ASSESSING TENDERS AND NOTIFYING RESULTS

The CPSSST recommends that all but the three lowest tenderers should be informed immediately that their tenders have been unsuccessful. The second and third lowest tenderers should be informed that their tenders were not the most favourable but that they might be approached again if it is decided to give further consideration to their offers. They should be notified at once when a decision to accept a tender has been taken.

Once the contract has been let every tenderer should be promptly supplied with a list of the tender prices. The estimator should record the results on the '**Comparison of results**' form.

11.02 ADJUSTMENT OF ERRORS

The estimator should be prepared to action any notification that the submitted tender contains errors in computation. Two alternative courses of action are possible:

(i) the tenderer will be given details of such errors and afforded the opportunity of confirming or withdrawing the offer. The estimator will need to refer to management when the extent of the computation errors have been determined for a decision as to whether to confirm the original tender figure or withdraw the tender. Under English Law it is possible to withdraw the tender at any time before its acceptance;

(ii) the second alternative is that the tenderer is given an opportunity of confirming his offer or of amending it to correct genuine errors. If the contractor elects to amend his offer and the revised tender is no longer the lowest the second lowest tender will be examined in more detail.

The estimator must consult with management to establish whether to amend the tender figure or to confirm the original offer, once the extent of the computation error has been determined.

In both situations the estimator or signatory to the original tender must be prepared to endorse the appropriate tender documents to note the acceptance or change to the tender. Such amendments should also be endorsed by the employer in the event that a contract is subsequently awarded.

PROJECT		Reference	
		Date	

COMPARISON OF RESULTS

Tender list	Tender figure		Contract period	Comment
	Fixed	Fluct.		

ANALYSIS OF RESULT

	Analysis of tender		%
1	Main contractor's work	Labour	
2		Plant	
3		Material	
4	Domestic sub-contractors		
5	PC sums nominated sub-contractors		
6	PC sums nominated suppliers		
7	Provisional sums and contingencies		
8	Project overheads		
9	Firm price/fluctuating price		
10	Other		
11	General overheads and profit		
			100

Comparison of results	£	%
This company tender		
Lowest tender		
Difference		

COMMENT

11.03 ACTION WITH A SUCCESSFUL TENDER

When a tender has been successful it will be necessary to take certain action before the actual contract documents can be signed. This action will be carried out by various people within the company's organisation but it is recommended that it should be co-ordinated by the estimator who was responsible for the original submission; the documents must be checked thoroughly from all aspects (technical, legal etc).

Contract documents

Contract documents should be checked to establish that:

- the drawings are those which were circulated with the tender documents;
- the dates, penalties and particulars etc given in the appendix to the conditions of contract are those stated in the tender document;
- the bills have been copied correctly from those originally submitted.

It is recommended that these comparisons of the contract documents are recorded on the adjudication report.

If there has been any delay in acceptance, enquiries should be made to ensure that there have been no changes which might necessitate reconsideration of the tender.

Distribution of information

The following information, as appropriate, should be made available immediately to those who will be responsible for construction and purchasing:

- site visit report;
- 'all in rate' calculations;
- calculation of project overheads;
- tender construction programme and method statement;
- report made by management;
- adjudication report;
- quotations received, noting those used in the tender;
- further information received after tender submission (if any);
- drawings, schedules, bills of quantity (priced and unpriced) and other relevant details.

It is recommended that an internal pre-contract meeting be held with all persons concerned in attendance so that details are fully discussed and the distribution of the foregoing information can take place at the meeting. At this pre-contract meeting a detailed report and discussion of decisions made at the time of estimating concerning methods of construction, site layout, organisation, suppliers and sub-contractors. These decisions should never be made binding on those to be responsible for construction. It is important that the reasons underlying such decisions and choices should be explained fully. However, decisions made at the time of tender should not be ignored unless it can be demonstrated that a better method of operation is possible.

Cost information

It is also recommended that a fully operational cost control system be introduced on all successful contracts. Although it is outside the scope of this Code it should be noted that the collection and collation of data is a specialised function involving the use of statistical and work study techniques. It is important that during construction information should be obtained on the labour and plant cost of major items or stages of work, on the quantities of material used, and on the cost of attendance on sub-contractors. This information is essential for control purposes and for guidance in future estimating. These records should be accurate and give all details including the details of the work, the position on the site of the operation, gang sizes, levels of incentives and bonuses being paid, plant, access, weather conditions, etc and include reference to all prevailing conditions in order that they may be interpreted correctly for future use.

Visit to the site

Those who were responsible for preparing the estimate should make periodic visits to the site during the progress of construction in order to:

- assess objectively the soundness of decisions made during the estimating process or subsequent site decisions where changes were considered necessary to improve performance and/or outputs;

- maintain a constant awareness of the conditions which prevail on sites.

Final reconciliation

At the completion of a project, the estimate and tender should be reconciled with the final cost and final account and the reasons for the results assessed in detail for guidance in future estimating.

11.04 ACTION WITH AN UNSUCCESSFUL TENDER

When the result of a tender is known the comparison of results form should be completed, where possible with a full list of the tenders and comparison made with the lowest (or the accepted) tender. An internal company memorandum should be circulated to all parties involved giving outline details of the tender results. Suppliers and sub-contractors who submitted quotations should be notified of the results as promptly as possible.

All tender documents should be carefully registered and filed for easy reference for possible future use if required.

ADDITIONAL READING

1 ELLIOTT, D. A. Tender patterns and evaluation in *The practice of estimating*. 1981. CIOB.